THE PROUST EFFECT

THE PROUST EFFECT

THE SENSES AS DOORWAYS TO LOST MEMORIES

CRETIEN VAN CAMPEN

TRANSLATED BY

JULIAN ROSS

OXFORD
UNIVERSITY PRESS

OXFORD
UNIVERSITY PRESS

Great Clarendon Street, Oxford, OX2 6DP,
United Kingdom

Oxford University Press is a department of the University of Oxford. It furthers the University's objective of excellence in research, scholarship, and education by publishing worldwide. Oxford is a registered trade mark of Oxford University Press in the UK and in certain other countries

First Edition published in 2014

Impression: 1

Published in the United States of America by Oxford University Press
198 Madison Avenue, New York, NY 10016, United States of America

British Library Cataloguing in Publication Data
Data available

Library of Congress Control Number: 2013948634

ISBN 978–0–19–968587–5

Printed in Great Britain by
Butler Tanner & Dennis Ltd

Contents

1

A memory of the senses

Introduction

Which of us is not familiar with those special moments, when we are taken by surprise by a tiny sensory stimulus that evokes an intense and emotional memory of an episode from our childhood? The scent of our mother's soap that takes us back to the familiar bathroom in the house where we grew up; or a song that takes us back to an emotional moment in our adolescent years. Once, returning home from a trip abroad, I was struck by an unknown yet familiar smell in the bathroom. The odour puzzled me; it didn't smell especially nice or unpleasant, that wasn't the point. Yet there was something intriguing about it. It was only after a while, perhaps after my next return from a trip, that I recognized the odour as being that of the bathroom in the stately waterfront house where my grandparents lived more than 40 years ago. At the moment of recognition I was transported in time and felt once again like the 4- or 5-year-old boy that I was in that house. I could see in my imagination the walls of the house, the water and the garden where we played, the faces of my then still young parents and grandparents. All sorts of sensory impressions came alive, drove the smell to the background and for a moment filled the decor of my youth.

Another sense memory that I share with many of my contemporaries in the Netherlands is the taste, and also the texture, of a kind of porridge named *Brinta* which is served at breakfast in many Dutch homes. When my own children arrived at the age for eating *Brinta*, I remember that just stirring the oats into the hot milk was sufficient to transport me back to the cold winters of the 1960s, reliving several memories of eating porridge that was too sticky (awful), too dry (almost impossible to swallow), and just the right creamy consistency, topped with my favourite crunchy layer of chocolate sprinkles (simply blissful).

And just one final sense memory from my early adolescence. Since the year 2000 a special non-stop, week-long radio show, playing a Top 2000 of the best songs ever according to listeners, has become a national event of collective nostalgia in the Netherlands between Christmas and New Year's Eve. After the glam rock era of the 1970s, with bands like Sweet, Mud, and Slade, who my classmates and I imitated in primary school with wild gestures and much bravura, an emotional episode in my life began with the song 'Sailing' by Rod Stewart. Just hearing a few notes, or his sandpaper-like voice, is enough to carry me back to the days of the ninth-grade school parties and my first experiences of dancing cheek to cheek to slow music, or 'grinding' as we called it somewhat graphically. I can see the faces of my classmates in the heavily saturated green and orange glow of the home-made disco lights, I can smell the smells and feel the textures of smooth and spotty cheeks, and relive the tingling feelings in my legs and the dozens of still unknown exotic butterflies in my stomach.

The smallest sensory impressions seem to slip through the cracks of my memory to bring back events that seemed long forgotten, and which now appear, large and real, in my consciousness. Not only do images appear, but sooner or later also sounds, smells, tastes, and the movements I made, the feeling of being touched by other people, sometimes even the details of the colour of a collar that someone was wearing or that tiny, irreverent wisp of the eyebrow lying horizontal above someone's eyes. Not only do I remember being there, but I also *feel* being there, even with a sense of contracting muscle fibres. Just like in those days, emotions bubble up through my stomach and seize my thoughts.

About a century ago, the French novelist Marcel Proust (1871–1922) described similar memories in his novel cycle *In Search of Lost Time (À la recherche du temps perdu)*. A famous anecdote from the cycle recounts how the taste of a madeleine cake dipped in lime-flower tea takes the main character Marcel back to his childhood, with all its colours, smells, and feelings.

This blissful experience inspired him to begin a search for his past. He devoted a great part of his life capturing the past in writing a cycle of seven books. In the last book he concluded that the sense memories were capable of affecting his deepest emotions and offering him deep insights into who he was and why he became a creative writer.

Reading Proust's reflections on the senses and memories, I wondered whether people in general have a 'memory of the senses'. Do the senses take information from special 'drawers' of memory? Do they have access to their own archives of tastes, smells, images, and sounds? What is this kind of memory that spontaneously calls up childhood memories without needing to be told? Why does this memory provide access to forgotten events in the past—events that we overlook when we actively search our memories?

We often speak about our memories as if we were talking about a computer memory. We 'save information' and 'open files' from our past. These mental acts are very useful in daily life when people ask us for information. They are even essential for knowing where and who we are. Think of people with dementia who lose their grip on their lives because they cannot remember who they are and what they just did. Apart from a 'computer memory' in the brain, however, people also have a body that receives impressions, such as a touch or a kiss, that we remember later with pleasure. Later we not only 'know' that we were touched, but we also 'feel' how we were touched and even what emotions it aroused. Why should we remember the feel of that touch? And why do these impressions of the past remain in the corners of our memory and catch us at unexpected moments with impressive emotional force, like the memory of an adolescent love affair? [1]

What is known about the corporeal aspects of remembering? The answer is, not much in comparison to the science of how the brain 'recalls' and 'forgets' information. Much research is focused on the deficiencies of memory, such as forgetting information, the (unconscious) distortion of memories, and the unreliability of testimonies. Every year, many scientific publications appear on the subject of deficits of memory and the recall of erroneous information. There are long-term research programmes focusing on subjects like dementia and false testimonies in court cases. And the scientific trail has been followed by countless self-help books that focus on solving memory deficits and improving memory.

In sharp contrast, little is known about the physical-aesthetical pleasures of remembering, like the blissful feelings that the simple taste of a madeleine cake evoked in Marcel Proust. For a long time, the anecdote of Proust's experience with the madeleine cake seemed to have been stored away as if in the curiosities department of a museum—fun to look at now and again but of little use in daily life. Fortunately, however, a number of artists and scientists eventually dusted off the madeleine anecdote and began to investigate this and other sense memories. The sciences and the arts can both provide insights into the field of the still relatively unexplored sense memories. The lifelong studies by Proust himself are an excellent example of this. Proust came from a family of physicians. He was expected to become a doctor himself, but finally found his destiny as a writer. This discipline led him to insights into the dynamics of the memory of the senses. While scientists tried to analyse memory into single elements, ready to be tested in laboratories, Proust and other artists searched for the meanings of memories in the stories people tell and the whole gamut of corporeal feelings and emotions they evoke in them.

Artistic and scientific studies are not always as strictly segregated as many people think. Scientists and artists have learned new things about the memory of the senses not only by experimenting with sense memories in their laboratories and studios, but even more so by participating in society.

To present one example at this early juncture, researchers and artists have learned from visiting older residents of nursing homes that they ought to be concerned not only with helping older persons to reduce their memory deficits, but should also devote attention to developing their powers of positive sensory reminiscences. People with dementia who have lost so much of their quality of life can experience moments of joy in sense memories. Little smells and tastes from their youth often bring ease and repose in their disrupted mental life.

But for people without mental problems, too, bringing back the past with friends by listening to music or watching pictures and films with larger groups of people during commemorations reinforces their sense of well-being and their social bonds. In art therapies, sense memories are actively evoked through forms of artistic expression like painting, looking at paintings together, making music and listening together, and writing autobiographical stories. Remembering can become a pleasant and meaningful activity again for people who are troubled by their past. It can make people relive their childhood and endow them with new insights into their personal lives.

Practical experiments and projects of this nature carried out by artists and scientists gradually lay bare the wondrous functioning of sense memories and enable us to begin to understand what sense memories do to people.

I have used three types of source for this book: scientific studies, art projects, and personal accounts. I have arranged the information in three parts. Part I explores what sense memories are; part II provides insights into their neurological, psychological, and sociological nature; and part III discusses the benefits and uses of sense memories for individuals and professionals.

In the first part of this book I explore what is so special about sense memories. The literary descriptions of the Proust effect by Marcel Proust have inspired not only writers but also aromatherapists, cooks, food designers, smell artists, art educators, and health care professionals. In addition to smell and taste, other stimuli such as sounds and images can evoke sense memories. For instance, songs from 'my generation' can elicit strong emotional memories many decades later, while from medieval painting to contemporary video art, images have been used to elicit special memories.

PART I

Art

2

The Proust effect

The dual effect of opening lost memories and bringing joy

As a child, Marcel Proust was a frequent visitor to the small town of Illiers, in Normandy, where his mother grew up. Illiers later served as a model for the town of Combray in Proust's magnum opus, the seven-volume novel *In Search of Lost Time* (*À la recherche du temps perdu*).

Nowadays, the little market town of Illiers draws many tourists. Today's visitor to Illiers will find all manner of tourist souvenirs and commemorative plaques on buildings, such as that adorning the premises of 'the butcher where Proust's mother bought her meat', probably in an attempt to evoke something of Proust's atmospheric descriptions in the visitor's memory. However, this is not the effect Proust had in mind.

As a young man, Proust lived in Paris, where his father was a professor of medicine, a career that he also had in mind for his sons.[1] Marcel's younger brother Robert fulfilled his father's wishes and did indeed become a renowned physician (Figure 2.1). Marcel, too, studied medicine for a while, but his asthma attacks ultimately forced him to abandon his training. In those days, asthma was regarded as a 'neurasthenia'—what today would be called a psychosomatic disorder. He was handled with kid gloves by his parents from early childhood, and lived a quiet life. Spending such a lot of time at home, he developed a special bond with his mother, whom he would later describe in loving terms in *In Search of Lost Time*. Her death therefore not surprisingly came as a terrible shock to Proust, who later declared that, 'When dying, Mom took with her the little Marcel'.[2]

After studying at the École des Sciences Politiques, Marcel began working as a librarian; that involved spending the day in a dusty library, and it is

Figure 2.1. Portrait of the French novelist Marcel Proust and his brother Robert as children. 1870s © UIG/images.de.

no surprise that as an asthma sufferer he was rarely seen there. He withdrew to his home and began writing more intensively. Friends recounted that he led a withdrawn life, sometimes even staying in bed for days on end. He received courses of treatment for his asthma from various doctors. In conversations with one doctor, Paul Sollier, he talked about 'emotional revivals' (*reviviscences*), a technique which Sollier saw as a way of opening up his patients to enable emotional problems from their past to be treated.

In the company of others, however, Marcel was a popular guest and entertaining conversationalist. He regularly interrupted his writing to immerse himself in the Parisian social life of *soirées*, a life which is painted in great detail in his books. Together with Oscar Wilde and fellow French novelist Joris-Karl Huysmans, he gave literary form to dandyism.[3] Proust epitomized both in practice and in writing what would in today's philosophical terms be described as the 'art of life'.[4]

In Search of Lost Time can be read in many different ways: as a biography, as a soap opera in which the same characters constantly reappear in a succession of love stories and tragedies, as a contemporary portrait of elite *fin-de-siécle* Paris society, and as a potent study of the power of human memory. It is this latter reading that I use for my study.

The first volume in the novel series, *Swann's Way* (*Du côté de chez Swann*), published in 1913, begins with the main character lying in bed: 'For a very long time I was in the habit of going to bed early'. In the hazy moments between waking and sleeping, his childhood and the world of his mother pass him by. As the story progresses, the young protagonist, who is also called Marcel, grows up, experiences his first female love, and later discovers his homosexuality. Winning and losing in love is a process that takes place against a backdrop of the Parisian *bourgeoisie*, with its *soirées*, country weekends spent at places like Illiers, and foreign travel to visit museums in Italy and Holland.

In the final volume of the series, *Time Regained* (*Le Temps retrouvé*), the main protagonist realizes that the hitherto opulent lifestyle is a thing of the past. The First World War has changed the world for ever. This might seem to be a sad ending, but Marcel has developed a memory technique for 'winning back' lost time. He observes that his senses allow him to 'rediscover' time that had been lost, in a process in which the past engages in a symbiotic relationship with the present and reveals new insights into his life. These sense memories provide the creative sources on which he is able to draw in his writing. In doing so, he takes the reader back to the beginning of the novel cycle, with the main character once again lying in bed and recalling memories of his youth.

A flavour memory that unfolds like Japanese paper flowers

In *Swann's Way*, the main character Marcel has little to occupy him and is bored, until one afternoon when he drinks tea with his mother. He receives an unexpected shock: something strange happens when he tastes a madeleine cake that he has dipped in his tea:

> Many years had elapsed during which nothing of Combray, save what was comprised in the theatre and the drama of my going to bed there, had any existence for me, when one day in winter, as I came home, my mother, seeing that I was cold, offered me some tea, a thing I did not ordinarily take. I declined at first, and then, for no particular reason, changed my mind. She sent out for one of those short, plump little cakes called 'petites madeleines,' which look as though they had been moulded in the fluted scallop of a pilgrim's shell. And soon, mechanically, weary after a dull day with the prospect of a depressing morrow, I raised to my lips a spoonful of the tea in which I had soaked a morsel of the cake. No sooner had the warm liquid, and the crumbs with it, touched my palate than a shudder ran through my whole body, and I stopped, intent upon the extraordinary changes that were taking place.[5]

In this absent-minded state, Marcel finds himself overcome by memories. It is an unusual but not unpleasant experience. Elsewhere in the book he compares it to seeing a wonderful painting in a museum or gallery for the first time, or contemplating a sublime landscape. He searches for the words to describe the indescribable:

> An exquisite pleasure had invaded my senses, but individual, detached, with no suggestion of its origin. And at once the vicissitudes of life had become indifferent to me, its disasters innocuous, its brevity illusory—this new sensation having had on me the effect which love has of filling me with a precious essence; or rather this essence was not in me, it was myself. I had ceased now to feel mediocre, accidental, mortal.[6]

The feeling disappears as quickly as it came, however. And when Marcel tries to recapture the experience by taking further sips of the tea, the experience appears to weaken as he continues to drink. In some surprise, he wonders whether the secret lies in the taste of the cake. To put this to the test, he takes a further bite of the tea-soaked cake and then arrives at a new insight:

> Whence could it have come to me, this all-powerful joy? I was conscious that it was connected with the taste of tea and cake, but that it infinitely transcended those savours, could not, indeed, be of the same nature as theirs. Whence did it come? What did it signify? How could I seize upon and define it?
>
> I drink a second mouthful, in which I find nothing more than in the first, a third, which gives me rather less than the second. It is time to stop; the potion is losing its magic. It is plain that the object of my quest, the truth, lies not in the cup but in myself.[7]

At that moment, Marcel becomes aware that a simple taste stimulus is able to evoke an aesthetic experience which goes far beyond the pleasant taste of the cake. The special qualities lie not in the cake, but in what it has awakened in him. The aesthetic experience is not limited to the moment itself, but extends far back into his past, calling up memories from long ago.

Although Proust's experience has often been portrayed in the popular media as a flashback, in fact the memory penetrated Marcel's consciousness only gradually. First there was the taste; this was followed by a vague feeling, and only then did sensory experiences from the past begin to appear in the form of remembered tastes, sounds, and images. Some time passed before the sensory impressions came together to form a clear memory and the penny dropped; it was only then that Marcel 'knew' what he was remembering:

> And suddenly the memory returns. The taste was that of the little crumb of madeleine which on Sunday mornings at Combray (because on those mornings I did not go out before church-time), when I went to say good day to her in her bedroom, my aunt Léonie used to give me, dipping it first in her own cup of real or of lime-flower tea.[8]

It was not the image, but the taste of the madeleine cake which created a link with the madeleines he had eaten as a child with his aunt Léonie. Seeing the cakes in shop windows had never elicited these memories in Marcel's mind:

> The sight of the little madeleine had recalled nothing to my mind before I tasted it; perhaps because I had so often seen such things in the interval, without tasting them, on the trays in pastry-cooks' windows.[9]

Marcel is amazed that such a tiny detail can evoke such an enormous memory. He becomes aware of the interplay of the senses in the construction of memory, the way in which the stimulus provided by the taste leads to the involvement of other senses. He represents this sensory interplay using a playful metaphor of the Japanese paper flowers which unfold in water, just as his sensory past has unfolded in his mind:

> And just as the Japanese amuse themselves by filling a porcelain bowl with water and steeping in it little crumbs of paper which until then are without character or form, but, the moment they become wet, stretch themselves and bend, take on colour and distinctive shape, become flowers or houses or people, permanent and recognisable, so in that moment all the flowers in our garden and in M. Swann's park, and the water-lilies on the Vivonne and the good folk of the village and their little dwellings and the parish church and the whole of Combray and of its surroundings, taking their proper shapes and growing solid, sprang into being, town and gardens alike, from my cup of tea.[10]

A touch memory of a Venetian pavement

Proust quickly realized that these surprising sensory memories were not limited to the domains of smell and taste; the 'madeleine experience' was repeated later with other sensory stimuli. Once again the incident takes place while Marcel's thoughts are elsewhere, this time suddenly forcing him to step out of the way of an oncoming vehicle. He stumbles, but quickly forgets his anger about the incident when the physical sensation of his stumble arouses memories of a pleasurable day spent with his mother in Venice. More quickly than with the madeleine cakes, all the sensory pleasures of that day are brought completely back to life.

This small impression of touch felt through his feet has aroused his other senses and transported Marcel from a French courtyard to the flagstones of St Mark's Square in Venice, where he had once spent time with his mother. Not only is he able to remember the incident, but also to feel it physically, and this sensation once again imbues him with feelings of joy and bliss:

> Reviewing the painful reflections of which I have just been speaking, I had entered the courtyard of the Guermantes' mansion and in my distraction I had not noticed an approaching carriage; at the call of the link-man I had barely time to draw quickly to one side, and in stepping backwards I stumbled against some unevenly placed paving stones behind which there was a coach-house. As I recovered myself, one of my feet stepped on a flagstone lower than the one next it. In that instant all my discouragement disappeared and I was possessed by the same felicity which at different moments of my life had given me the view of trees which seemed familiar to me during the drive round Balbec,

the view of the belfries of Martinville, the savour of the madeleine dipped in my tea and so many other sensations of which I have spoken.[11]

Eagerly searching for a way to experience the intense feeling of happiness again, Marcel, as with the madeleine cakes, tries to relive the memories by repeating the same sensory stimulus. Though it does not produce the result he is hoping for, it does open another doorway:

> Merely repeating the movement was useless; but if, oblivious of the Guermantes' reception, I succeeded in recapturing the sensation which accompanied the movement, again the intoxicating and elusive vision softly pervaded me as though it said 'Grasp me as I float by you, if you can, and try to solve the enigma of happiness I offer you.' And then, all at once, I recognised that Venice which my descriptive efforts and pretended snapshots of memory had failed to recall; the sensation I had once felt on two uneven slabs in the Baptistry of St. Mark had been given back to me and was linked with all the other sensations of that and other days which had lingered expectant in their place among the series of forgotten years from which a sudden chance had imperiously called them forth.[12]

A sound memory of a train wheel

Marcel's sense impressions in his present are not always the same as those he experienced in the past, as the tasting of the madeleine cakes and tripping over a paving stone were. The last example of a sense memory describes a sound memory in which the sounds that link present and past are quite different: a spoon clattering on a plate and the clang of a hammer on a train wheel:

> The servant in his ineffectual efforts not to make a noise had knocked a spoon against a plate. The same sort of felicity which the uneven paving-stones had given me invaded my being; this time my sensation was quite different, being that of great heat accompanied by the smell of smoke tempered by the fresh air of a surrounding forest and I realised that what appeared so pleasant was the identical group of trees I had found so tiresome to observe and describe when I was uncorking a bottle of beer in the railway carriage and, in a sort of bewilderment, I believed for the moment, until I had collected myself, so similar was the sound of the spoon against the plate to that of the hammer of a railway employee who was doing something to the wheel of the carriage while the train was at a standstill facing the group of trees, that I was now actually there.[13]

Once again, the sense memory appears at an unexpected moment, and it is as if the sense of hearing and memory have forged a pact to invade Marcel's consciousness. The other senses rapidly join in to overwhelm him, who makes no objection to this and allows himself to be transported to a woodland setting from his childhood years.

Time regained

The final fragment takes us to the last volume in the novel series, *Time Regained* (*Le Temps retrouvé*). In this volume we find long, reflective passages by the writer Marcel in which he considers the functioning of human memory. In describing Marcel's musings, Proust was not so much analysing the act of remembering or calling up factual knowledge—to use modern terminology, how information is stored in the brain and how it is recalled when needed—but was much more concerned with aspects that today receive little attention in neuropsychological research on memory, such as the sensitivity, the creativity, and the joy of remembering.

Inspired by his conversations with his doctor, Paul Sollier, and following in the footsteps of the philosopher Henri Bergson (1859–1941), whose lectures he had followed at the École des Sciences Politiques in Paris, Proust distinguished between two kinds of memory: voluntary and involuntary memory.[14] Voluntary memory is governed by the will of the individual and is goal-directed. It is used, for example, to recall who was present at a birthday party or who we met this morning at the baker's. By contrast, involuntary memory functions largely independently of personal will, breaking into consciousness unbidden and at unexpected moments. In the character Marcel's case, this mainly happens at moments when he is feeling slightly detached, daydreaming or absent-minded. At those moments he is evidently more receptive to the sensory stimuli which transport him back to his youth.

Proust believed that involuntary memories possessed an aesthetic quality. For him, the coinciding of aesthetic awareness in the present with an aesthetic awareness from the past was what made involuntary memories so special. The sensory stimulus and the memory of sensory images, sounds, tastes, and smells had an artistic quality and provided important material for Proust the artist.

Marcel's senses work in harness in this creative process. It is not only a taste from the past that comes into his consciousness, but also images, sounds, smells, tactile impressions—virtually the entire spectrum of sensory observations. And the experience is not so much one of a picture or photographic record of an event in the past, but rather of a collage of impressions. Proust distanced himself from the idea that the memory was a sort of photographic archive of images. He believed that different senses combined to create living observations and memories:

> A picture of life brings with it multiple and varied sensations. [...] What we call reality is a relation between those sensations and those memories which simultaneously encircle us—a relation which a cinematographic vision destroys because its form separates it from the truth to which it pretends to limit itself—that unique relation which the writer must discover in order that he may link two different states of being together for ever in a phrase.[15]

For Proust, the act of remembering is the act of *reliving* experiences from the past. Marcel relives the past with the body, the knowledge, and the emotions he has in the present and uses them to construct a new emotion, which contains characteristics of both youthful and adult emotions. Proust stressed that he was able to enjoy his memories so much because as a child he was receptive to the details in the world around him. Because he had derived such conscious pleasure from the taste of madeleines in his youth, for example, his memory was better able in later life to make connections with the past. As a writer, he rediscovered significant moments from his past thanks to the aesthetic impressions that he had collected as a child. And he saw art as a means of knowing this reality of significant memories:

> Then a new light arose in me, less brilliant indeed than the one that had made me perceive that a work of art is the only means of regaining lost time. And I understood that all the material of a literary work was in my past life, I understood that I had acquired it in the midst of frivolous amusements, in idleness, in tenderness and in pain, stored up by me without my divining its destination or even its survival, as the seed has in reserve all the ingredients which will nourish the plant.[16]

Proust died before he was able to complete his novel series, as a result of a neglected lung infection. He was 51. The final volume of his novel, *Time Regained*, was published 5 years after his death.

The work drew admiration in contemporary literary circles when it was published, initially because of its complexity and its psychological insights. But it was only midway through the twentieth century that it came to be recognized internationally as a masterpiece that had changed literature. Important writers in every language region of the world have been influenced by Proust's work.

More recently, neuroscientists and philosophers of mind have rediscovered Proust's original thoughts on autobiographical memory, as I shall discuss in Chapter 6. Before going into scientific theories on the dynamics of the Proust effect, I will first explore the phenomenon in a number of literary and artistic investigations.

3

The power of fragrances

Smell memories in literary fiction and culinaria

It was Proust who made the concept a familiar one to us, but what we call the Proust effect had in fact already figured in French literature since the beginning of the nineteenth century, as the French historian Alain Corbin demonstrates in his colourful study of smells, *The Foul and the Fragrant* (*Le miasme et la jonquille*).[1] The experience described by Proust was already known in French literature as 'odour memory'.[2] However, it is worth noting that the Proustian example of the madeleine cakes actually describes a flavour memory, whereas we often think of an odour memory today when speaking of the Proust effect.[3]

As early as the end of the eighteenth century, the French philosopher Pierre Maine de Biran wrote in his diary about 'this strange sensation, which swept away the veil between heart and thought, destroyed the distance separating past from present, and produced the melancholy of the "never more" through awareness of the unity of "I"'.[4] Maine de Biran had an original view of memory which, almost two centuries before memory psychologists arrived at this insight, distinguished between different memories for ideas, emotions, and habits.[5] Odour memories, he argued, were more closely associated with emotions than intellect and evoked nostalgic, melancholy feelings:

> Odours, linked to such spontaneous, ineffable feelings as are experienced in youth, always awaken more or less the same feeling; you find yourself young again, in love, in a scented shrubbery. It is there that the heart plays out its game independently of thought; when the veil is lifted, we feel all that we have lost, and melancholy seizes our soul.[6]

Nineteenth-century French literature contains many examples of smell memories. The French authoress George Sand is a case in point; in her autobiographical *Story of my life* (*Histoire de ma vie*), published in 1854, she describes the nostalgic pleasure aroused by the sweet scent of honey as it transports her back to her youth:

> So, seeing the convolvulus in flower, she [the mother] says to me: 'Smell them, they smell of good honey; and do not forget them!' This is therefore the first revelation of the sense of smell that I remember; and by a link between memories and sensations that everyone knows, and cannot explain, I never smell convolvulus flowers without seeing the place in the Spanish mountains and the wayside where I first plucked them.[7]

It was quite usual in the second half of the nineteenth century to employ nostalgic smell memories in novels. For many writers, smell memories were more than simply a recollection of an actual event from the past, but evoked a physical feeling, a confluence of emotions, which appeared to have greater meaning: a sign on the wall. The influential poet and essayist Charles Baudelaire established a link even then between what is today known as the Proust effect and synaesthesia.[8] His poem *Correspondences* (*Correspondances*), which would later be adopted as a programme by many symbolist painters, contains the following lines:

> *Comme des longs échos qui de loin se confondent*
> *Dans une ténèbreuse et profonde unité,*
> *Vaste comme la nuit et comme la clarté,*
> *Les parfums, les coleurs et les sons se*
> *répondent.*
> Like prolonged echoes mingling in the distance
> In a deep and tenebrous unity,
> Vast as the dark of night and as the light of day,
> Perfumes, sounds, and colors correspond.[9]

According to Baudelaire, 'the senses communicate with each other' ('Les parfums, les couleurs et les sens répondent'). The title of his poem, *Correspondences*, also refers to this phenomenon. The senses speak to each other and together evoke magical moments from a distant past ('prolonged echoes').

In another poem from the same *Flowers of Evil* collection, entitled *Perfume* (*Le parfum*), Baudelaire ecstatically declaims:

> *Charme profond, magique, don't nous grise*
> *Dans le présent le passé restauré!*
> O deepest magic charm's sweet thrall
> In present or in past restored![10]

Odour was given special status by Baudelaire and others because of its 'bewildering power to evoke the past' as Corbin put it.[11] And it will come as no surprise that when describing smells, poets and writers thought above all of the fragrance of loved ones. In a time when few if any photographs were available, scented objects offered a means of evoking memories of a loved one. Corbin illustrated this nicely with a number of excerpts from the correspondence between Gustave Flaubert and his lover Louise Colet. Not only did they exchange handwritten letters, but as time passed also added mementos such as handkerchiefs, gloves, and scented locks of hair. Flaubert describes this process in the following fragments taken from letters written in August and September 1846:

AUGUST 6: 'I look at your slippers, your handkerchief, your hair, your portrait. I reread your letters and breathe their musky perfume.'

AUGUST 9: 'I'll take another look at your slippers again...I think I love them as much as I do you...I breathe their perfume, they smell of verbena—and of you in a way that makes my heart swell.'

AUGUST 11: 'In daydream I live in the folds of your dress, in the fine curls of your hair. I have some of those here: how good they smell! If you knew how I think of your sweet voice—of your shoulders and their fragrance that I love.'

AUGUST 13: 'Your mitten is here. It smells sweet, making me feel that I am still breathing the perfume of your shoulder and the sweet warmth of your bare arms.'

AUGUST 15: 'Tell me if you use verbena; do you nor put it on your handkerchiefs? Put some on your slip. But no—do not use perfume, the best perfume is yourself, your own fragrance.'

AUGUST 27–28: 'Thank you for the little orange blossom. Your whole letter smells pleasantly of it.'

AUGUST 31: 'Thank you again for the little orange blossom. Your letters are perfumed with them.'

SEPTEMBER 20: 'A thousand kisses...on those long curl papers; I sometimes breathe a little of their odour in the small slipper with the blue slashes, because it is there that I have packed away the lock of hair; the mitten is in the other one, next to the medal and beside the letters.'[12]

These examples make clear that, in nineteenth-century literary circles, smell memory was more than just a mere pleasant memory. Smells evoked mixed emotions of childhood happiness, melancholy about time lost, sometimes leading to sadness-inducing nostalgia.

What was the purpose of these restless emotional experiences? Baudelaire described them as signs on the wall.[13] Proust embarked on a self-examination and discovered that sense memories did not simply penetrate into his life and disrupt his equilibrium, but actually provided him with an insight into unprocessed emotions.[14]

Odours are able to penetrate through to the emotional layers in the subconscious realms of the human psyche. In modern literature, the book *Perfume* (*Das Parfum*) (1985) by the German writer Patrick Süskind is a story of smells and smell memories which swirl around in the psychopathology of a serial murderer.[15] The book recounts the story of a boy from the eighteenth-century Paris slums. His exceptional sense of smell, and above all his ability to analyse smells immediately into their constituent ingredients, enables him to escape from his lowly background and rapidly forge a successful career for himself in the world of the Paris *parfumiers*, and not long afterwards to move on to the centre of the French perfume industry in the town of Grasse. Smell memories and his youth appear to play virtually no role, until he evolves into a murderer.

The main character himself has no scent, and murders a series of young women in order to obtain an extract of their scent from their skin. His reason for doing this is because of his dejection at what he has lost. And, this being a romantic novel, what he has lost is of course love. As a youth, he had a very fleeting experience of love with a young woman in Paris, and ever since he has been searching obsessively for the feminine fragrance which enveloped her. He is searching for the fragrance of lost love. As a psychopath, he murders young women for their bodily scent. As a maker of perfumes, he is able to distil the 'scent of love' from their dead bodies. Ultimately, the murderer is arrested and convicted, but just before his hanging, he manages to mislead the officials and the crowd by drenching himself in this self-concocted scent of love. The bystanders are enveloped in a reverie of amorous memories and the prisoner escapes—only to be killed later, back in the poverty-stricken slums of his birthplace.

At a deeper level, the narrative is about the emotions, memories, and desires that are awakened by smells. Sadly, the main character meets his end before arriving at a place of self-knowledge. The reader, by contrast, is privy to that insight. Smell memories which evoke mixed emotions can be harbingers of the re-emergence of an emotional past with which the individual has not yet come to terms.

Smell and colour in nursing homes

The power of sense memories has not only been used in the literary genre, but in other artistic disciplines too. A group of visual artists recently began designing and constructing installations, mostly in museums and galleries, which allow visitors to experience at their leisure taste and olfactory stimuli that have the power to activate subconscious sense memories.

The Flemish artist Peter de Cupere, for example, designs smell installations where visitors can activate and deactivate smells in order to envelop themselves in either a sweet-smelling or malodorous atmosphere.[16] For a nursing home in the Dutch town of Doetinchem, he designed three 'Smelloflowers' made from plastic-coated metal (see Figure 3.1), one for each of the three residential pavilions for older persons with dementia,

Figure 3.1. 'Smelloflowers' in the Sensire Den Ooiman care centre in Doetinchem (the Netherlands) by artist Peter De Cupere, <http://www.peterdecupere.net/ 2013> (with permission from Peter De Cupere). © Peter De Cuper, 2013. See also plate 1.

and each with its own distinct colour and scent. Each pavilion thus has its own 'departmental flower': the 'Hyagreencinth', the 'Tulorangeip', and the 'Magbluenolia'. Each flower is imbued with its own scent, which wafts upwards from the calyx of the flower. The colours are repeated throughout the rooms in framed prints depicting the 'departmental flower'. The effect is to create orange, green, and blue areas in the nursing home.

The scents of the 'departmental flowers' can change in character with the revolving seasons. For example, the 'Hyagreencinths' give off the scent of pines around the Christmas period; another flower wafts the smell of turkey around the pavilion, while the third spreads the aroma of pancakes. When spring comes around, the flowers give off the fresh scent of spring flowers. The flowers can also be used for short periods to administer aromatherapies, for example. For people with dementia, their senses and memories of their youth are often all they have left to hold on to in their daily lives.[17]

The Dutch food designer Marije Vogelzang carries out artistic projects in which the central focus is on the design and meanings of food. For a project in Rotterdam about memories of the Second World War bombardment of the city in 1940, for example, she designed an exhibition

Figure 3.2. Wartime food by eating designer Marije Vogelzang, <http:www.marijevogelzang.nl> (with permission from Marije Vogelzang). © Marije Vogelzang, 2013. See also plate 2.

with dishes taken from the period, when the city was confronted with severe food shortages. The exhibition included small dishes of 'mock meat' ('Valsch vlees'), concocted from beans and flour (see Figure 3.2). This elicited both positive and negative reactions from visitors who had experienced the War as children, and who remembered not only the hunger, but also the pleasure they felt when they did have something to eat, and the sense of bonding and togetherness they experienced whilst sharing a meal from the meagre rations.[18]

Tasting rooms

'Taste artists' like Marije Vogelzang are thin on the ground, but there are many chefs and restaurateurs who are experimenting with the interplay between the senses and the memory. Restaurants can be experimental tasting rooms. Restaurateurs, of course, understand better than anyone that people taste food not only with their tongue and nose, but also with their eyes, ears and the tactile sensors on their lips and in their mouths.

Tastes interact with other sensory impressions, and with memories, and subsequently embed themselves in the memory, creating a complete picture together with other impressions such as images of the restaurant interior, the sounds of cutlery clattering against porcelain, the background music, the voices of the diners and the shouting waiters, and the emotions of the evening (dining with a new lover or with old friends). One stores all of these impressions and then buries them deep in the memory of the senses, where they remain dormant until that same taste or a related taste reawakens them years later.

Tastes, like smells, have the peculiar property that they are difficult to extract from the memory through conscious effort. We often know that we tasted or smelt something for the first time somewhere, but are unable to recall or re-experience the taste memory at will. This veiling of taste and smell impressions by our sense memory takes place with enormous speed. In a restaurant, for example, when the waiter asks me if I have enjoyed my meal I know whether or not it was good, but I can no longer recall precisely how it tasted. The taste impression has already been stored deep in my memory. If I later experience the same taste, everything comes to the surface again: the faces of my tablemates, the interior of the restaurant, the background noise, the voices, the emotions, and so on. Accessing these memories is achieved much more quickly and accurately via sensory stimuli than through words.

A psychological experiment involving taste shows that our taste memories do not consist of pure tastes, but rather of mixed sensations which include seeing, hearing, touching, and above all smelling. When we taste something, other sensory perceptions also play a role. During a conference in Oxford, I once took part in an experiment hosted by top chef Heston Blumenthal of the Fat Duck Restaurant and the famous psychologist Charles Spence from the University of Oxford.[19] All those present were served with a portion of ice cream which had the unusual flavour of bacon and eggs. We were also given two questionnaires and asked to fill them in whilst eating the ice cream. Both questionnaires contained the same question: does this ice cream taste more like eggs or more like fried bacon? Alert to the tricks that those hosting such experiments sometimes like to play, I soon noticed that they were playing sounds in the room: first the sound of clucking hens, followed by the sound of bacon sizzling in a frying pan. But even though I was aware of this, it seemed to me whilst hearing the sound of chickens that the ice cream tasted more like eggs, and that whilst hearing the frying bacon it tasted more like bacon. And it made no difference how much I tried to fight against that perception.

Anyone wishing to see for themselves how visual impressions influence taste might like to take part in one of the various 'dinner events' such as *Dinner in the dark*, where, as the name suggests, participants are served a meal in total darkness.[20] Although touch and hearing help in our perception of taste, the lack of visual input changes the way the dishes taste. I myself once took part in a blind wine-tasting. As a wine lover, I thought I wouldn't be easily fooled. The task I was given was a simple one: I was blindfolded and my tablemate passed me a cool wine glass which I put to my lips using only my sense of touch. She wrote down my extravagant characterization of the taste of the wine and finally noted whether I thought the wine was red or white. In most cases, I got that latter question wrong. The leaders of the experiment had served the red and white wines at the same temperature, and to my consternation I discovered that, when blindfolded, I wasn't even able to tell the difference between white wine and red wine.

Eating is about more than just tasting. Take the animated comedy film *Ratatouille*, for example.[21] The main character is Remy, a rat with a fondness for cooking and a very special sense of taste: he experiences tastes in the form of shapes and music. By chance he ends up working in the kitchen of a Paris restaurant, which develops after his arrival into the most popular establishment in the French culinary world. Though of course it is essential that no one must ever know that there is a rat working in the kitchen. His counterpart is a grumpy, misanthropic restaurant critic, for whom having rats in the kitchen would naturally be anathema. Ultimately, however, Remy wins over the critic by serving him a dish which on tasting the first mouthful catapults him to his youth and the taste of the ratatouille made by his mother in the kitchen at home. At that moment, this blissful memory is associated for the critic with the tasting of the dish in the restaurant. This genuine Proust effect transforms the cynical critic into a relaxed *bon viveur*. The film thus illustrates in a playful and natural way how the Proust effect can completely overturn someone's attitude and view of life.

The film contains an assortment of scenes in which tastes recall other sensory impressions: from the melodies and abstract images in the main character's sense of smell to the blissful childhood emotions of the restaurant critic. Tastes and smells trigger visual and musical impressions to create Proust effects. Actually, images and music have that same power, though they work differently. The accounts in Chapter 4 show how music has the wondrous power to bore deep into someone's personal past with apparently superficial melodies.

4

Listening to 'my generation'

Musical memories by pop songs

For Marcel, the main protagonist in *In Search of Lost Time* (*À la recherche du temps perdu*), hearing the music of Vinteuil—a French composer invented by Proust the author—opens the door to the same kind of joyful memories as the taste of the madeleine cakes had done (see Chapter 2). Vinteuil's music brought together all the sensory impressions:

> [...] In that instant all my discouragement disappeared and I was possessed by the same felicity which at different moments of my life had given me the view of trees which seemed familiar to me during the drive round Balbec, the view of the belfries of Martinville, the savour of the madeleine dipped in my tea and so many other sensations of which I have spoken and which Vinteuil's last works had seemed to synthesise.[1]

Today, pop songs are the vehicles which evoke most childhood memories in adults. The Japanese writer Haruki Murakami named a book about a teenage love after the song *Norwegian Wood* that the main character hears as muzak in an aeroplane whilst coming into land. As his plane approaches touch-down, the 37-year-old protagonist listens to this 'calming' Beatles song as a means of reducing the tension he feels whilst the plane is landing, but finds that hearing the song makes him unwell. He suddenly starts to smell wet grass, feels the wind on his skin, and hears the sound of birds. He is carried back 18 years and relives his life as a student aged nearly 20:

> Memory is a funny thing. When I was in the scene I hardly paid it any attention. I never stopped to think of it as something that would make a lasting impression, certainly never imagined that 18 years later I would recall it in such detail. I didn't give a damn about the scenery that day. I was thinking about myself. I was thinking about the beautiful girl walking next to me. I was

> thinking about the two of us together, and then about myself again. I was at that age, that time of life when every sight, every feeling, every thought came back, like a boomerang, to me. And worse, I was in love. Love with complications. Scenery was the last thing on my mind.
>
> Now, though, that meadow scene is the first thing that comes back to me. The smell of the grass, the faint chill of the wind, the line of the hills, the barking of a dog: these are the first things, and they come with absolute clarity. I feel as if I can reach out and trace them with a fingertip. And yet, as clear as the scene may be, no one is in it. No one. Naoko is not there, and neither am I. Where could we have disappeared to? How could such a thing have happened? Everything that seemed so important back then—Naoko, and the self I was then, and the world I had then: where could they have all gone? It's true, I can't even bring back her face—not straight away, at least. All I'm left holding is a background, pure scenery, with no people at the front.[2]

Murakami was in fact writing a variation of Proust's madeleine anecdote. In his main character, too, emotions overshadow the sounds of the song and the memory hits. The other senses join in and reach into his inner being to retrieve the past emotions. Coming back little by little, the memories of these past emotions return: his love for a mentally confused girl. The more clearly he remembers the physical details, the stronger the emotion becomes.

Norwegian Wood is about coming to terms with loss. In his teenage years, the main character first loses his childhood best friend, who unexpectedly commits suicide, and then later his closest childhood girlfriend who, probably as a result of the loss of that same friend, gradually loses her grip on reality and ends up in a psychiatric institution in the mountains, where she too eventually takes her own life. At the end of the book, an old woman plays the song on guitar for him and, enveloped in the sound memories it evokes, he is able finally to come to terms with his loss. The circle is complete. The book which begins with a light-hearted sound memory of a pop song, closes with the same sound memory, but now charged with emotion and insight.

There is a peak period in life when people absorb memorable pop music: what people regard as the music of 'my generation' begins around the age of 14 and ends somewhere in their late twenties.[3] On average, people's strongest emotional memories of music date from the time around their sixteenth birthday.[4]

Lists of the top pop hits of all time are popular in several countries, not least because they allow listeners to give full rein to their sentimental and nostalgic feelings without embarrassment. The stories recounted by listeners are interesting because they sometimes reveal how sound memories work.

In the Netherlands, the massive public response to the annual 'Top 2000' list on Radio 2 is a good example of this. Many songs from the past evoke special memories in listeners, who on hearing a song are often transported back to that time in their lives. In 2006, letters from listeners were collected for the first time in the archive of *Geheugen van Nederland* ('Memory of the Netherlands').[5]

The past is relived in a plethora of sensory details which are bound together, as it were, by emotions felt in the past. Nostalgia is king. One of these sound memories is that of listener Janny van Eldik-Bonouvrié, who describes how hearing the song *Bernardine*, a hit for Pat Boone in 1957, transports her back to her life as a 4-year-old child:

> Our first memories are usually images, flashes. Later, we place them in the context of their time. If I count back, my first musical recollection dates from the year 1957. I had my fourth birthday in November of that year. [...] When it rained, the washing was dried indoors on washing lines in the kitchen and on one of those wooden racks that was placed around the black solid-fuel stove. Dad came home at five o'clock and we had a hot meal. We had a radio—tall, with a rounded top and two knobs, one for the volume and one for tuning.
>
> [...] The first song that wafted out of the loudspeaker was Bernardine by Pat Boone. I couldn't get enough of it. When the song started, my father turned the radio louder, swept me up and danced me around the room singing 'Oh, Bernardine' along with the record. Sometimes I would ask my father to put Bernardine on the radio; as a three year-old, you obviously don't know that you can't have records played on demand; you think your father can do anything, including playing Bernardine whenever you want to hear it.
>
> [...] I can still feel now what I felt then. I know now that it is a safe, happy feeling, being in a setting with the certainty that everything will remain the same. That kind of sense of security can carry you through your whole life and is something you would wish for every child. Thank you, dad; you're 83 years old now, and just as jolly as you always were. You still sing in a choir and you still know how to enjoy yourself.[6]

When hearing this song, the listener recalls among other things the swooping movements of the dance, and all those sensory recollections come together to create an overwhelming sense of safety.

Danny Gordijn has placed a different memory in the archive of the Top 2000. Hearing the 1973 song *Sebastian* by Cockney Rebel not only takes him back to the past, but also brings him to a reflection on memory. His account illustrates well how a sense memory is constructed. It is not a photographic recollection of an event in the past, but rather a collage of feelings that are associated with images, sounds, and other impressions, in the perception of a child with no sharp dividing lines between reality and fantasy:

> I have a terrible memory. [...] That makes it all the more surprising that I can still remember very clearly how my childhood mind responded to music. I still know exactly what a flood of images some songs aroused in my somewhat overly fanciful little head. [...]
>
> There's one image that will never let me go: I was on holiday with my family in Norway or Sweden or somewhere. I don't remember exactly where, of course, but from the back seat of the car I could see the endless mountains and fir trees passing by, with the occasional stream or lake. Somewhere during the trip through that beautiful Scandinavian landscape, I had my comic book out and was lying on the back seat. My father, who was an avid compiler of holiday tapes, switched on the cassette recorder. I heard several all-time favourites coming out of the machine, but also a song that until then I'd perhaps heard once or twice and which had a frighteningly big effect on me—it was a song that I was as afraid of as I was awestruck by: Sebastian by Cockney Rebel. My English wasn't too good at the time—come on, I was five years old! Or six, or eight, I can't remember exactly. But I lost myself in the hypnotic sounds of that irresistible voice. Somewhere between sleeping and waking, in a majestic Scandinavian setting, the song created an image in my mind that has never left me since: in my mind's eye I saw grey-hooded monks, dragging heavy chains and coming together in an underground cave. In their midst stood a blonde, terribly afraid young boy (no, not necessarily me). The atmosphere was threatening, but with the possibility of escape. When the monks call the boy Sebastian, he can leave, escape from the cave. That realisation was as inexplicable as it was rock-solid: when someone calls him Sebastian, he is safe. I know, it's not exactly a healthy childhood fantasy, but that's how it felt at the time.[7]

Once it has become associated with a particular moment, a pop song can stay with you for the rest of your life. The following account, from a listener called Herman, is a lament that is associated with the song *The Green, Green Grass of Home* by Tom Jones:

> She was six and so was I. We were in the first year at school, a church school. She came up to me in the school playground and said: I'm going with you. I received postcards from her. I can't remember if I sent cards back.
>
> When we were 15, I saw her cycling past our house, carrying her ice-skates with her. Her clothes were bright blue and green. It was in the Christmas holidays, and the radio was playing 'The Green, Green Grass of Home'. I grabbed my skates and cycled after her, but I wasn't able to find her.
>
> I thought I was in love, but I was too shy to do anything about it. I thought that if you were in love, it was always reciprocal. I also thought that if you

thought about somebody that you're in love with, they thought about you at the same time. I had very romantic ideas.

One day I took my courage into my hands and went to the grocers where she worked, to buy an ice cream. My heart stood still and I hardly dared say anything when I was stood face to face with her, and she said even less. Sometimes I saw her in the church; sometimes I even managed to cycle close behind her when she left the church to go home. Unfortunately, she only lived a couple of hundred metres from the church.. .

Later, I left the village and didn't see her any more. But she kept on cropping up in my dreams, and she still does to this day, 30 years later. Those dreams always take place in the house where she lived, by the railway bridge. She must be somewhere in that house, but I never manage to see her. Even today, whenever I pass that house on the train—something I do often—I look to see whether she is perhaps standing by the railway bridge. I still ask myself whether my feelings were reciprocated.[8]

Special teenage memories acquire a soundtrack, as it were, as illustrated in the account by listener Douwe Gerlof Heeringa. His tough time working as a planter in the searing Australian heat takes on a nostalgic tone when he hears the song *Pride* by U2.

I think it was about 37 degrees and, judging by the shrinking stack of lettuce plant trays on the flatbed of the Holden Utility, we would be heading back to the hostel in Griffith, New South Wales, Australia within a quarter of an hour. I'd been Down Under for around three months and had worked all that time. As a 19 year-old backpacker, you take on anything that comes your way. Same with this job: planting cultivated lettuces in the iron-red soil in a field in the middle of nowhere.

The alarm clock went off at 5:00 a.m. every day. And every day at 5:45 a.m. the boss was waiting for us outside in his car. [...] At around 2.00 p.m. every afternoon we were driven back, tired, sweaty but with a sense of satisfaction at another day done. Waiting for us was the shower in the former school that had been converted to an international hostel. This was home to a bunch of backpackers who helped out various farmers in the area during the harvest season. At about half past two, when it was too hot to work on the land, the Brits, Canadians, Dutch and Irish all made their way slowly in their work clothes from the kitchen to the bedroom.

Some preferred a quick shower, while others liked to relax with something cold out of the fridge. Not the Irish, though: every day without fail, at around 2:30, they would start playing a CD really loud in their dormitory. Then they would go to take a shower further down the corridor. This particular CD was by a band that I wasn't familiar with. But just like the routine of smearing on sun lotion, taking a coffee break, driving through the fields, during

> those weeks I became familiar with the guitar riffs of this 1980s band. I could sing the words 'One man came in the name of love' without faltering while washing my hair. Ultimately of course I asked the Irish guys what the CD was. It turned out to be a compilation album by the Irish band U2, 'Best of 1980–1990'.
>
> Around five weeks later, after something like 50,000 lettuce plants had been put in the ground and the work was finished, I left for Kangaroo Island. It was to be my first trip, and the promised highlights were 'sleeping under the stars', 'wildlife' and 'barbecues'. A small group of us spent three days roaring around the island, with two cassette tapes blaring in the four-wheel-drive: one was a collection of Ozzie Rock and the other was 'U2 1980–1990'. That was the start of my love for U2. The connection with Pride involves the smell of shower gel, rows of work boots, red sand on long corridors and kangaroos in the night, and rocky coasts with here and there a sea lion.[9]

Personal memories evoke emotions. So does listening to music with someone else, as the earlier account shows, when the pop song *Pride* evoked a shared memory of coming home to the hostel in Australia at the end of the working day.

Music memories are different from smell and taste memories because they also have a social function. Smell memories are often intimate and personal, whereas music memories are often shared with peers. Everyone has their own personal *Pride*, which is connected to a place and time from their youth.

The shared memories may also relate to a particular time: the 'my generation' feeling that songs can evoke—whether it be The Beatles or The Rolling Stones, Pink Floyd or Fleetwood Mac, Joy Division or The Cure, Nirvana or The White Stripes, to pick just a few at random—of successive generations of young people, who have now become older people.

Listening to music with others has changed in recent decades because of technology used for playing music. Until the 1980s, young people took vinyl records up to attic rooms to gather around a turntable and listen to new music. Following the emergence of the compact disc, they were able to bring, and swap, larger numbers of albums.

More recently, the distribution and sharing of music via the Internet has changed the way we listen to music.[10] Initially somewhat doubtful in terms of its legality, sharing new music has now become a legitimate part of the social media landscape. Social media such as LastFM, Flickr, Myspace, and Facebook offer new ways of calling up musical memories, a fact illustrated by the makers of *Pensieve*. *Pensieve* is a web application that is named after a memory stone from the Harry Potter series.[11] The software supports everyday reminiscence by emailing memory triggers to people that contain either social media content they previously created on third-party websites or text prompts about common life experiences. It was developed at Cornell University in Ithaca in the United States by the information scientist Dan Cosley and his staff. Participants are sent an e-mail each day containing a song, photograph, video clip, or something else that they themselves once posted on a social medium such as Facebook. In each e-mail, or on their personal page on a social medium, they receive a sensory stimulus at unexpected moments and open a diary in which they can write down a memory, which they can then choose to share or not. The e-mail containing the 'memory trigger' arrives unexpectedly, and the content is also a surprise. The participants then write a blog describing the memories that have been triggered in them. A small survey linked to the experiment showed that participants' sense of well-being increased and that they more consciously 'came to terms with their past'. Or, as one participant put it, 'I try to connect it to issues I am dealing with currently. I think, like therapy, this helps me understand myself'.[12]

By delivering the sensory stimulus at unexpected moments, *Pensieve* approaches the Proust effect. The trigger is unexpected, is sensory in nature, and evokes involuntary memories. *Pensieve* could be thought of as a sort of 'artificial Proust effect', or a translation of the Proust effect into a contemporary means of communication (for example, via social media). It stimulates the personal and social awareness of social media users, whose actions and above all thoughts and memories are often guided by the stimuli posted on these platforms by other users. *Pensieve* is just one example of the web tools that are available for triggering reminiscences; others include Caraclock and the Personal Digital Historian Project.[13] These systems display similarities with projects such as the 'Storytables' which encourage older residents of care homes to recall and share memories (as we shall see in Chapter 9). These residents do not naturally use social media such as Facebook, although I would not rule out the possibility that the next generation of residents may well do so. It is then possible that old and young will find themselves sharing memories together on Facebook, or whichever social medium is 'in' at that time. Finally, remembering is an intrinsically social endeavour, originally learned with parents and developed with friends throughout the person's lifespan.[14]

As digital stimuli, sounds and images are becoming ever more interwoven in our consciousness of the world around us. As memorabilia, however, images and music have different histories. Music touches deep emotional layers in the brain; images seem to trigger more of a factual recollection of how things were at a particular time, but they also affect people emotionally. Images come in many forms, from drawings to digitally manipulated visual environments. How images have stimulated memories from the Middle Ages to modern-day video and Internet is the subject of Chapter 5.

5

The art of memory

Visual memories through paintings, television, and video art

For centuries, mnemonics, or the art of memory, was an essential technique for conveying information. The best-known memory technique is that developed by the Ancient Greeks, who used the 'method of loci' (places). The story goes that the Greek poet Simonides of Ceos discovered the method by chance when he was a guest at a wedding but was called away after having read out a poem. After he had departed, the roof of the banqueting hall caught fire and collapsed, killing all the guests. The victims were so disfigured that their relatives were unable to identify them. Simonides was able to do so, however; he recalled the names and faces of those present by comparing the place where the victims were found with his memory of how they were seated around the table. In doing so, Simonides demonstrated that the spatial arrangement of images, associated with persons or objects, can be a powerful mnemonic tool.

Until the invention of printing, stories were recounted using images (think of the linear stories depicted in medieval frescoes on Italian church walls) and songs (performed by troubadours). The amount of available space (the canvas, the length of the song) was literally too limited to recount stories in great detail, as was possible later in books. The images and songs served as triggers for the memory to enable people to recollect past events.

Painting

The painters of the late Middle Ages and early Renaissance period were masters in 'placing' important items (symbols) and people (saints) in their paintings. A painted landscape took on a unified character and each unique background contained niches where these symbols or saints could be localized. In her book *The Art of Memory*, the British art historian Frances Yates shows, for example, how the Italian painter Giotto di Bondone (*c.*1267–1337) demonstrated his mastery of this technique.[1] In the Arena Chapel in Padua, he subtly varied the backgrounds of the symbols used to represent virtues and vices (for example, Charity and Envy, see Figure 5.1), enabling the loci to be fixed in the memory of the viewer. Giotto is considered to be one of the first painters to apply the technique of linear perspective in art. According to Yates, Giotto was searching for artistic techniques to give his figures space and therefore a locus (place) in the composition.

In Figure 5.1, the artist's use of perspective means that the virtues appear to step forward out of the painted niches, enabling them to be inscribed more recognizably and more sharply in the memory of the viewer. In this way, Giotto literally created space for the loci.

In later years, books took over from paintings as mnemonic vehicles, but for a long time images retained their influence as memory aids. Dante

Figure 5.1. Fresco paintings by Giotto showing the virtue Charity (on the left) and the vice Envy (on the right) in the chapel at Padua.

Alighieri's (*c.*1265–1321) literary sketch of the multi-layered Hell in the *Divine Comedy* is an example of an application of the method of loci, which is still very expressive and representative of the art of its time. The Italian writer Ludovico Dolce commented as early as 1562 that 'Dante's *Inferno* can be regarded as a memory system designed to enable Hell and its punishments to be fixed in the memory through the use of striking images in carefully arranged positions'.[2]

A fresco by Domenico di Michelino (Figure 5.2) depicts Dante holding a copy of the *Divine Comedy*, pointing to a procession of sinners being led to Hell. Behind him we see the seven levels of Hell, with at the top the earthly paradise of Adam and Eve. On the ascending and descending circular pathways, Dante positioned the events of his story in such a way that it could be remembered as a pathway of loci.

Leonardo da Vinci (1452–1519) went a step further. Not only did a painting serve to help the viewer remember events, but Da Vinci also wanted to create a suggestion that the viewer was 'physically' present at

Figure 5.2. Fresco of Dante and his Divine Comedy by Domenico di Michelino, circa 1465 in Florence Cathedral.

the depicted event, with all senses engaged.[3] The linear perspective was another technique used to convey this visual impression of being present; Da Vinci used this technique in his most famous biblical mural, *The Last Supper*, in the church of Santa Maria delle Grazie in Milan. In this painting, Da Vinci combined the method of loci, which he used to position Jesus and the Apostles at the table, with the linear perspective which draws the viewer in and gives them the impression of physically standing before the table and witnessing in person the moment at which Jesus bids farewell to his disciples. Using these artistic techniques, Leonardo constructed an intense experience: a viewer standing in the right position in the refectory has the feeling of being present today at an event which took place long ago.

Playing with memory images

We are overwhelmed with so many images today that we rarely view them with any sense of awe and dedication. Visual artists have understood this and therefore play with everyday images such as photos, TV images, and home videos to interfere with our observation and intensify our memories. The American artist Bill Viola (1951) has been experimenting for several decades with everyday images such as home videos. In one of his most recent projects he uses video images to elicit reactions from the public which in their intensity resemble the reactions of the Renaissance public to the dramatic paintings by artists such as Michelangelo or Pontormo. Viola makes videos of the facial expressions of people in mourning, for example, or when meeting up with old friends. He shows these videos on flatscreen monitors which are the size of paintings and are hung on the walls of museums and galleries. The images of the facial expressions are slowed down to the point where they are barely moving. Through the use of colour, lighting, and above all the baroque expressiveness (for example, Michangelo's *Pietà*, with the arm dangling beside Christ's body), the effect is to evoke strong associations with Italian paintings from the sixteenth century.

Bill Viola's video *The Greeting* is on permanent display in the De Pont Museum of Contemporary Art in Tilburg, the Netherlands. It has been exhibited in many museums and galleries throughout the world. Inspired by a painting by Jacopo Pontormo, Viola shows a video lasting around 10 minutes in which three women appear dressed in period clothing (Figure 5.3).

Figure 5.3. Left: *The Visitation*, 1528–1529, by Jacopo Pontormo (Church of San Michele in Carmignano, near Florence). Right: a still from the video *The Greeting* (1995) by Bill Viola (De Pont Museum, Tilburg). Video/sound installation, 430 × 660 × 780 cm. Colour video projection on large vertical screen mounted on wall in darkened space; amplified stereo sound. Performers: Angela Black, Suzanne Peters, Bonnie Snyder. Production still. Photo: Kira Perov. See also plate 3.

The recording time of just 45 seconds is slowed down to a playback time of 10 minutes, causing the gestures and facial expressions to become almost completely still, as if in a painting. The images are accompanied by the noise of a loud, rushing wind, which is mimed in the images to some extent in the flapping of the women's clothes.

The video begins with two women in conversation. The facial expression of the older woman suddenly changes when a third woman joins them, though she is still out of frame. This is followed by a hearty greeting, which takes the form of slow movements in which every expression, every emotion appears to be exposed in the utmost detail. As a result, the viewer sees much more than they would at normal playback speed. The slowness also seems to make the emotions more intense. The slowed-down gestures and facial expressions evoke memories of the viewer's own life, but above all, of the Renaissance and the mannerist paintings of Pontormo.

Time appears to stand still in these emotional moments. As a viewer, we lose track of time. When I visited the exhibition, I sat in a darkened room with a constant rushing sound in my ears, which already confused my sense of time. The video images, which provided the only light in the dark room, also offered no sense of time, because they were running even more slowly than the usual slow motion on TV, making it impossible to judge the speed of the actions. Does the embrace of the two women last for a couple of seconds or half a minute? Recollections of paintings and personal events disrupt one's fragile sense of time even more, cutting through it with moments taken from other times and other places. At the end of the video you awaken from a daydream as if you had drifted off to sleep, with no sense of whether you have sat there for a few seconds or several minutes. The text at the end of the film tells you that it was 10 minutes.

Time appears to pass more slowly during important emotional events. We commonly hear people say: 'Time stood still'. Proust had already noticed that when past and present coincide in a sense memory, time briefly appears to stand still. By slowing down time in his video, Viola may be creating such a moment whilst at the same time giving the viewer an opportunity to observe the emotions of the moment—a hand gesture, the casting down of the eyes, a fluttering lock of hair—more closely and more intensely.[4]

Tactile television

The videos of Bill Viola not only show that the image is merely a detail of the memory, but also make clear how the other senses, especially touch, are brought into play. Image and touch are closely linked in human perception. The British philosopher George Berkeley noted as long ago as the eighteenth century that optical images only become real when our sense of touch is brought to bear.[5] Optically, a ball looks like a flat, two-dimensional circle. It is only by touching it that we understand its spherical shape, according to Berkeley. This idea was given a surprising twist a few centuries later in the reaction of philosophers and artists to the emergence of the new medium of the twentieth century: television. The Spanish artist Salvador Dali, with his predilection for images that connected with childhood memories from the subconscious mind, expressed a desire for 'tactile cinema' even before the arrival of television. He wanted every film scene to be tactile, right down to the last details: 'A character touches a corpse and your fingers sink into the putty,' wrote Dali in a letter to the filmmaker Luis Buñuel.[6]

Their ideal would be given form in television, according to the media expert Marshall McLuhan, known for coining the phrase 'the medium is the message' in his book *Understanding Media*.[7] McLuhan situated the emergence of television in the history of the mass media—and, I would venture to add, in the history of the (memory) techniques—as the next step on from the introduction of the printing press. While printing had replaced the oral Western culture with a visual culture, McLuhan saw the global penetration of television as the beginning of a culture in which the balance between the senses would be restored.[8] McLuhan understood that television did not give viewers the feeling that they were looking at pictures of an event, as with a photo or painting, but rather the feeling that they were actually physically present at the event being depicted. Television images gave the illusion that they could be touched. McLuhan went a step further than Leonardo da Vinci; the viewer not only feels physically present, but also has the feeling of being able to reach out and touch the scene portrayed.

Memory manipulation

The fact that memories consist not only of sounds and images, such as archive film images, but thanks to their interaction with the senses are interwoven with other sensory impressions such as smell, taste, and touch, is nicely demonstrated—probably unintentionally—in a well-known film about memories. The film, *Eternal Sunshine of the Spotless Mind* by Charlie Kaufman and Michel Gondry, produced in 2004, tells the story of a young man and woman who fall in love and embark on a relationship.[9] When the feelings fade, the woman decides to attend a special clinic to have all memories of her former lover erased from her mind. All items that remind her of him are collected together and shown to her one by one, while her brain patterns are scanned at the same time. All the articles are then removed and, while she is sleeping at night, the scanned brain activity patterns are identified in her memory and eradicated. On discovering that she has done this, the man later undergoes the same procedure himself.

The erasure of memories and emotions is presented in the film as if it were a normal scientific technique.[10] With quasi-scientific thoroughness, all images and sounds which recall the lover and the relationship are erased from the patient's memory. However, the therapist erases only visual (photos, souvenirs, gifts) and aural (songs) stimuli from the patient's brain, but not the smell, taste, and touch impressions. When the former lovers meet each other again by chance, their mutual scents, tactile impressions (handshake, a fleeting touch), and smell and touch impressions (a first kiss on the cheek) evoke deeper emotional memories involving smells, tastes, and touch. The upshot is that the main characters fall in love all over again.

There are a number of magical moments in the film when the two characters do not recognize each other (they know each other, but their faces have been erased from their memories), but feel very comfortable with each other: 'I don't know you, but I have the feeling that I've known you for a very long time'. This is a reference to the fact that, in love at first sight, lovers often have the feeling that they have known each other for a very long time. And, the film suggests, perhaps that is indeed the case, and their memories of each other have simply been erased at some time in the past.

New media

Sound and image recordings mean that people today have access to far more images and sounds to use as tools for accessing past memories, such as film clips showing the first steps taken by a young child, cassette tapes recording their first words, cinema newsreels during the opening of a new town hall, television recordings of sporting events, or radio journals containing the sounds of a flood disaster.[11]

Today's impressions of events are often digitally recorded and there would seem to be virtually no need for us to recall events spontaneously, using our own powers of imagination. On the other hand, people have more ways of accessing their memory with the help of the senses. Anyone who has walked through a museum for television and radio, for example, knows that they can be almost overwhelmed by Proust effects. Often it is not so much the content as the type of media; think of the sound of an early gramophone or the grainy images of a black and white television that trigger the memories.

Since the television has come to dominate the way we organize our living rooms, many domestic scenes have become associated with a television that is switched on. In a museum showing television programmes, the visitor may see a programme from the 1970s out of the corner of their eye, and almost immediately they feel the atmosphere of that time.

What effect will new media have on our personal and collective memories a few decades hence?[12] Will our sense memories become more superficial now that they are there for the taking, as it were, in recorded images and sounds? Will the past still be brought to life in smells and colours in the near future? Will Proust effects still be able to move us emotionally then?

In the past, visions of future technology have usually proved to be exaggerated, as science fiction stories from past decades demonstrate. On the other hand, technology will progress further in the digitization of sensory stimuli; in addition to digital images and sound recordings, people in the future may also be able to record and replay smells, tastes, and tactile impressions digitally in new storage media.[13]

Proust effects will still occur in the digital age. Even with new digital stimuli, sensory memories are still reconstructions of the past.[14] The physical reliving of a personal history remains the essence, whether the stimuli that provoke this are natural or digital. The stimuli will change, but people, their bodies, emotions, and brains that relive sense memories will remain the same.

PART II

Science

6

The hippocampus of Proust

The making of sense memories in the brain

Sense memories follow wondrous pathways. How does the brain create sense memories? Which pathways through the brain do sense memories follow? How are the other senses engaged, for example, on a smell memory pathway to someone's personal past? After becoming acquainted in Part I with the variety and power of sense memories in various artistic practices, in Part II we will explore the science of the human brain in search of neurological and psychological explanations for the wondrous pathways of sense memories.

Neuroscientists have begun showing a great deal of interest in the Proust effect in the last decade; this is striking, given that the phenomenon lay more or less ignored by memory scientists for almost a century, or was even denied and banished to the category of 'literary fantasies'. Scientific researchers appear to have rediscovered the Proust effect.[1]

Analysing and testing the Proust effect in research laboratories, as well as interviewing individuals about it in surveys, have produced surprising insights. Within the wide realm of sense memories with which we became acquainted in the foregoing chapters, I will now focus on the Proust effect. In order to avoid confusion, I define the Proust effect here as an involuntary, sensory-induced, vivid and emotional reliving of events from the past.[2]

Smell memories

Although it took almost a century before scientists rediscovered the Proust effect, the first scientific study of this phenomenon actually took place shortly after the publication of Proust's oeuvre. In a psychological study conducted at Colgate University in New York in 1935, 254 male and female subjects were asked whether they had ever found that smells could evoke intense and living childhood memories.[3] A majority were able to recount such experiences. Women did so more often than men: one in five men were able to recall a smell memory, while nine out of ten women were able to do so. In addition to the first simple statistics, the stories of smell memories provided an insight into how memories are gradually formed after a meaningful smell reaches the nose. Donald Laird devoted an article to these accounts entitled 'What can you do with your nose?'. One woman, for example, had the following story to tell:

> One afternoon at the age of twenty-five while reading a, book a, very strange sense of *loneliness* came over me. Various memories of my childhood came up before me, especially the memory of books that I had read as a child. I looked the book over very carefully and could find no reason for this sudden recall. The print of the book had a strange pleasant odor. I turned to the front of the book and found that it had been printed in London. I have since found that English and American print have very different odors. Several times since then I have had this same experience and when turning to the front of the book have found that it was printed in London. As a child my books all came from England. Odors play a very important role in my life. Under nervous strain I have found that perfume brings relaxation and at night will put me to sleep.[4]

More recently, the Dutch psychologist Douwe Draaisma from the University of Groningen also devoted himself to an analysis of the stories recounted by people about their memories.[5] Draaisma told me about two people from his immediate setting who had each revealed an interesting characteristic of the Proust effect. The earlier account by the American reader and the following two anecdotes from Draaisma make clear that Proust effects do not occur in a flash, but actually enter the individual's consciousness very gradually:

> When I was writing the chapter on smell memory in my book, a colleague told me the following story. He grew up in the village of Dinteloord, in the southwest of the Netherlands, but has now lived in Groningen for around forty years. He had a dream that he was back in his childhood bedroom in Dinteloord. He woke from his dream, which had seemed so real that he was briefly disorientated and thought he really was waking up in his childhood bedroom. He lay there thinking about it, wondering where such an intense childhood dream had come from.
>
> Suddenly he understood what was happening. The harvesting and processing of the sugar beet (known as the 'sugar beet campaign') had started in

> Groningen that night. There are only a few places in the Netherlands which have sugar beet factories, and they include Dinteloord and Groningen. The smell of the sugar beet processing had evidently created an association in his subconscious mind. It was only later that the conscious memory came to him, so that he was able to understand how he had ended up in that dream. For me, this was a typical story; it often works like that: first a sort of mood or feeling is aroused by a particular smell, and it is only later that the memory which is responsible for it comes to the fore.[6]

The second anecdote is about a woman from the village of Haren, south of Groningen, who told her story to Draaisma after he had given a talk:

> In early September she often felt a bit sad and didn't understand why, because there was nothing special happening—until she realised that the smell of the sugar beets being roasted outside smelt the same as during the war. The sugar beet campaign always starts in early September in Groningen. Here again, she was first aware of a mood, and it was only later that the memory came out.[7]

These three accounts show that sense memories often begin with a mood, an undefined feeling which the person finds difficult to place. It is only later that they understand where the feeling is coming from, are able to give it a specific context of sensory impressions, and can experience a past event anew.

Smell experiments

Accounts such as these offer excellent insights, but are less useful for testing scientific hypotheses about the special way in which the memory of the senses functions. Are sensory stimuli able to elicit different recollections from words? Laboratory experiments are needed to determine this. Half a century after the study in New York, the American memory psychologist David Rubin and his colleagues at Duke University in Durham, North Carolina attempted to test experimentally a number of assumptions about the Proust effect. He presented a group of students with 15 familiar odours[8] in three forms: as odours they could smell, as images representing the odours, and as written words describing the smells. The participants in the study were asked to rate the memories that were triggered by the different stimuli in terms of how real they appeared and how pleasurable they were. The statistical analysis of the descriptions did not however reveal any significant differences between the reactions to the actual smells, the illustrations, or the verbal descriptions. The only difference was that the real smells more often triggered a memory that the subject thought had been lost forever. The demythologizing conclusion of the researchers was that Proust effects are no more real or pleasurable than other memories.

A number of reservations can be made about this conclusion, however. For example, the students were asked to call up memories, and therefore made use of what Proust called the *voluntary memory*. According to Proust, however, his experiences came from the *involuntary memory*.[9] Moreover, the students in the experiment had to react quickly to the stimulus and locate their experience on a dimension, something that is at odds with Proust's description (and the anecdotes of Laird and Draaisma) of the experience as a gradual process in which the memory unfolds step by step.

At the end of the 1990s, the British psychologists Simon Chu and John Joseph Downes from the University of Liverpool discovered that Proust effects are tied to age. They visited a group of older persons with an average age of 70 years and investigated their smell memories.[10] The subjects were presented with two types of stimulus—actual smells and smell-related words—and asked to describe the memories triggered by these stimuli. Later they noted down the approximate year that the remembered events had taken place. The analyses showed that smelling the actual smells triggered older memories than reading smell-related words: the smells evoked the most memories from when the subjects were aged between 6 and 10 years, whereas the words brought back the most memories from the period when they were aged between 11 and 25 years.

This sheds new light on the so-called 'bump' in the lifespan of memories. In another study, Rubin and his colleagues found that when older people are asked verbally to recollect their past, they report more memories from their twenties and thirties than other decades in their life.[11] Later, this bump was also found in involuntary autobiographical memories of older people.[12]

One would expect involuntary memories to have an earlier 'reminiscence bump', as suggested by the study by Chu and Downes. The Danish memory researcher Dorte Berntsen from the University of Aarhus has extensively investigated the nature and development of involuntary memories in participants who kept diaries.[13] She found no difference: involuntary autobiographical memories show the same distribution across the person's lifespan as voluntary memories. In both cases verbal cues were used, which may explain why the same distribution was found.

Other researchers have argued that emotional childhood memories, and especially traumatic ones, are difficult to access using verbal and linguistic cues such as stories and diaries, but can be opened using situational cues such as perceptions and sensations of the environment at that time.[14]

After having reviewed these studies, my conclusion is that the 'bump' in childhood sense memories lies somewhere between the ages of 3 and 10 years.[15] This earlier 'bump' of sense memories is supported by an interesting explanation. In common voluntary verbal memories, adults describe the events in their lives in rich narrative terms. Using just a few words or lines gives them access to the narrative structures that they built in their childhood. However, their time travel along these narrative constructions are limited by the fact that these narrative constructions are incomplete or incoherent at the age of 8 years, and begin to become broadly coherent at the age of 12, with this coherence increasing further as adolescence progresses.[16] Before the age of 12, young children are still developing their storytelling abilities, and their narratives are consequently neither complete nor coherent. This can be readily tested by asking an 8-year-old and a 14-year-old to tell a story about an event at which they were both present. The story of the 8-year-old will contain more sensory impressions, loosely joined together, while the story by the 14-year-old will be more abstract, contain less sensory detail, and have a stronger narrative structure. Between the ages of 7 and 11 years, changes take place in the length and complexity of children's autobiographical narratives.[17] Children aged between 10 and 12 years produce narratives that more effectively engage the listener with the time and place of the event, and they sustain and elaborate on topics more effectively than children aged between 7 and 9 years.[18] Yet even at age 11 to 12 years, children's narratives still lack the causal connections that characterize older adolescents' and adults' narratives.[19] It is only in adolescence that individuals construct an extended life narrative.[20]

Later in life, experiences that were stored in stronger and richer narrative structures will be more easily retrieved using verbal questions and by telling stories. By contrast, the less fully narrated experiences with rich sensory details will be more readily accessed through smells, tastes, sounds, images, and touch.

Why do we lose our early childhood memories? For most adults, the age of earliest memory is typically around 3 to 3½ years, though there are pronounced differences between individual adults.[21] The number of memories from the first 7 years of life is smaller than would be expected due to normal forgetting alone, implying an accelerated rate of forgetting memories from early childhood. At least until the age of 11 years, children forget childhood memories at an accelerated rate.[22] Scientists explain this by the fact that children do not have the full ability to store their experiences in narrative structures that can be accessed in adulthood.[23] Another possibility that has barely been explored in scientific studies is that these presumed 'lost' memories can be more readily found when adults use the path of sense memories, triggered by situational and sensory details.

Are sense memories different from verbal memories? Rachel Herz and her colleagues at Brown University in Boston, Massachusetts, also reported on a series of experiments in the early 1990s which showed that smell memories are non-lingual. They asked subjects to smell different odours and describe the memories they triggered. In a third of cases the subjects were unable to name the smell. This suggests that the memory was not elicited by using words but rather via a sensory pathway.[24]

Proust wrote that he had regularly seen madeleine cakes at the baker's, but that those images had not triggered such strong emotional childhood memories as the taste and smell of the cake that he dipped in his tea. Herz and her colleagues investigated this phenomenon in one of their next experiments. They offered their subjects alternating smells and images and asked them to say which evoked the stronger emotional memories. An American participant, for example, recounted that she remembered a trip to Paris when she was aged around 10. She described her mother putting on her makeup to go out and applying *Opium* perfume. In her session, this scent and a photo of the perfume bottle were used as stimuli. Both from the reports by the participants and from scans of their brains, it transpired that the smells triggered stronger emotional memories than the images.[25]

In yet another experiment, Herz compared the power of smells, images, and sounds to trigger emotional autobiographical memories. She presented the subjects with campfire, new-mown grass, and popcorn as stimuli in three sensory forms: as an image, as a sound, and as a smell. The memories triggered by the smells were found to be more emotional and more evocative than those elicited by images or sounds associated with those stimuli. However, there were no significant differences in the level of detail or vividness of the memories.[26]

Smell memories, then, are more evocative and more emotional than memories elicited by images and sounds. Could this be because the olfactory centres in the human brain are located close to the emotional centres? In asking this question we are embarking on a new terrain, namely that of brain research.

Brain pathways

How do our brains process a Proust effect? Are there neurological explanations for Proust's observations, for example, that sense memories are involuntary, trigger an emotional, aesthetic experience and construct the past creatively?[27]

Proust wrote that his alter ego Marcel was often in an absent-minded state at the moment that the intense, emotional childhood memories broke into his consciousness. Neurological research has shown that this phenomenon can be attributed to the actions of the 'implicit memory'. According to the American memory researcher Joseph LeDoux, emotional events leave traces in two different memory systems: on the one hand the conscious memory system which records the facts of events (who, what, where, when), and on the other in the memory system which records the emotions and impressions of the event in the subconscious.

LeDoux described these two systems as the *explicit* and *implicit* memory, respectively.[28] In the experiences described by Proust, the implicit memory awakened a vague feeling or emotion ('a wonderful feeling of joy') in his conscious mind. This vague feeling was only later recognized by the explicit memory which added place, time, and situation to the feeling and therefore made it concrete.[29] Several brain areas play a key role in this process.

The activity of the amygdalae, the parts of the brain which add emotional colour to experiences and memories by releasing hormones, are responsible for the sudden awakening of the undefined feelings of joy and happiness at the start of Marcel's memory. Later, the hippocampus adds substance to the memory by activating memory pathways to the parts of the brain that process the senses. Memories are widely distributed in the brain, and the hippocampus helps in recovering these memories. The hippocampus is essential for our spatial orientation, both when navigating our way through the physical world, when recalling past events, and when imagining future events.[30] (The Appendix contains a fuller description of the amygdalae, the hippocampus, the frontal lobe, and other parts of the brain that are involved with autobiographical memories.)

After the implicit memory has broken into Marcel's consciousness at an unexpected moment with an undefined feeling, his brain goes in search of the origin of the feeling. The physical impressions (tripping on a paving stone, the clattering of a spoon against a plate, the taste of a cake) activate other sensory perceptions via the hippocampus, such as images and sounds. From the activated memory pathways the frontal lobe of his brain, which acts as a sort of 'office manager' in the brain issuing search instructions, gathers information and makes a judgement as to what goes with what, creating a composition of the past which appears to be the most coherent and consistent. This creates the presentation of the past of which Marcel ultimately becomes aware.

The addition by the hippocampus of a specific sensory context to the undefined feeling retrieved by the amygdalae was demonstrated physiologically by a group of British neuroscientists at University College London, led by Jay Gottfried. Using brain scans, they found significant physiological indications for the Proust effect in the hippocampus.[31] Memories of an event are distributed across the sensory centres in the brain and are brought together by the hippocampus. The researchers found that when a sense memory occurred, activity could be detected in the hippocampus, the stimulated sensory domain, but also in other sensory domains.[32]

The hippocampus records memory pathways of events. Memory pathways run through different areas of the brain responsible for processing sensory information, emotions, and actions. A simple sensory stimulus can be enough for the hippocampus to activate the entire pathway.[33] This fact makes clearer why the simple taste of a madeleine cake triggered a world of sensory and emotional impressions in Marcel.

We can now also answer the question of why smells evoke more emotional childhood memories than other sensory stimuli. The brain scans of the participants in the experiments by Herz et al. described earlier showed that smell stimuli stimulate the amygdalae and the hippocampus more powerfully than images.[34] The answer to this question is therefore that the amygdalae are situated closer to the areas of the brain responsible for smell and taste than to the areas responsible for vision, hearing, and touch.[35] The amygdalae are in turn situated close to the hippocampus (see Figure 6.1). Put simply, the neural pathways are shorter and therefore faster. Smells and tastes therefore recall the past in a dynamic way more quickly thanks to the short distance to the hippocampus.

Creative memory

Proust described remembering as a creative act, as the creation of joyful moments in which present and past are brought together. Can this creative function of the Proust effect be localized anywhere in the brain? If the

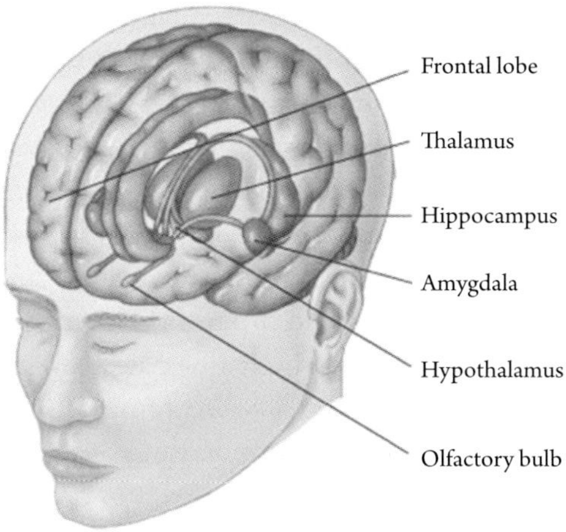

Figure 6.1. Locations of the hippocampus, the amygdala, and the olfactory bulb in the brain.

amygdalae are responsible for emotions and the hippocampus for multiple sensory perceptions, which area of the brain is able to create this sense of 'joyful beauty' from these ingredients? I did not find a neurological answer, but did at least encounter some philosophical answers.

The American psychologist Russell Epstein from the University of Pennsylvania compared Proust's ideas about the involuntary memory with those of his contemporary William James.[36] At the end of the nineteenth century, James formulated the psychological theory of the 'stream of consciousness', the continuous flow of thoughts and emotions through the conscious mind. The stream of consciousness comprises both core thoughts which appear clearly before our mind's eye and more marginal thoughts and feelings, which Williams described as the 'fringe'. These 'fringe' thoughts feel as if they are more on the periphery in a slumbering consciousness, but at any moment they can leap out from the wings into the spotlight of the mainstream consciousness. When experiencing sense memories, for example, elements pop into Marcel's consciousness which appear to be unrelated to each other. A spoon clattering against a plate, for example, awakens memories of the heat of summer, the smell of smoke, and images of rows of trees.

Following on from Proust, Epstein distinguished between a voluntary and involuntary memory, placing the latter in the 'fringe' category. Whereas the voluntary memory works in a targeted way, for example, retrieving a fact from the memory, the involuntary memory is less focused, seeking out all manner of connections without a particular purpose. However, describing this phenomenon as 'fringe' does not explain the creative power that Proust attached to sense memories.

The linguist Renate Bartsch from the University of Amsterdam proposed an explanation; she repeated that Proust did not see the memory as an archive from which items can be retrieved.[37] Rather, the memory 'recreates' the past by linking different recollections and constructing something new from them in the present. This created past is not a portrayal of the factual past, but a meaningful whole in which the original elements largely lose their individual meaning. I compare it with the perception of a *Gestalt*, as in Figure 6.2; a different configuration of the three elements results in a new image of a white triangle.

A memory is more of a construction than a reconstruction: memories never reconstruct the past precisely, but add elements from later and from the present, thereby creating a new experience with elements from different moments in a person's life.[38] A stimulus such as tasting a madeleine cake subconsciously activates memory pathways such as emotions, motor actions, cognitive content, and sensory perceptions. These largely subconscious processes in the fringe of the stream of consciousness connect this mental content and at a certain moment the creations come together in the full spotlight of the conscious mind. It was this which gave rise to the conscious experience of Marcel's mornings spent in Cambray with his Aunt Léonie.[39]

This makes clear that the Proust effect is not the recollection of a single moment, but a construction of different elements from the past. Memory researchers distinguish between different forms of autobiographical memories: (1) memories of periods, such as the 1970s or the primary school years; (2) memories of recurrent events, such as sporting events on Saturdays; and (3) memories of unique events, such as first experiences.[40] Only the latter

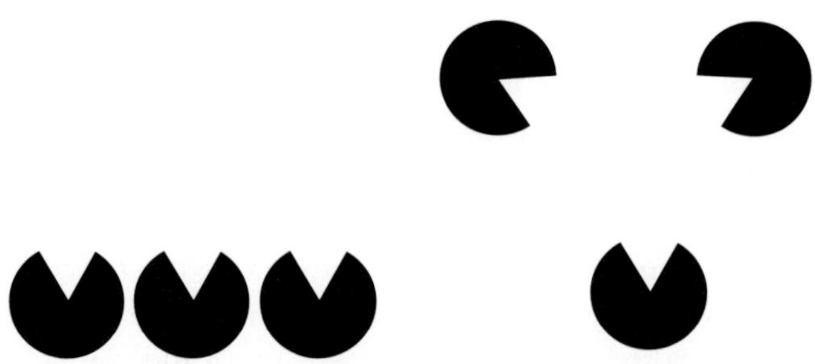

Figure 6.2. Elements separated and in configuration, resulting in a new Gestalt.

memories come close to a reconstruction of a single situation from the past; the first two variants are almost always amalgamations of elements from different moments.

Remembering is thus a process of construction. Sense memories closely resemble the creative process that is experienced by an artist. This helps explain why sense memories had an artistic function for Proust. In the final volume of *In Search of Lost Memories* he came to this conclusion:

> [...] whether it was a question of impressions given me by a view of the Martinville belfry or memories like those of the two uneven paving-stones or the taste of the madeleine, it was necessary to attempt to interpret them as symbols of so many laws and ideas, by trying to think, that is, by trying to educe my sensation from its obscurity and convert it into an intellectual equivalent. And what other means were open to me than the creation of a work of art?[41]

Proust's brain

It would be fascinating to see what was going on in Proust's brain. This raises the challenging question of what we might have been able to see if his brother Robert, the doctor, had had a brain scanner and had invited his brother Marcel to take his place in the apparatus to undergo a series of sensory stimuli which would trigger Proust effects in his memory. Or if Marcel's brain had been preserved after his death so that it would be possible to study its anatomy now. Would certain parts of his brain show differences from the norm?

We know from neurological research that parts of the brain that are heavily used are relatively larger than those that are used less. In a study of London taxi drivers, it was found that knowledge of the London street layout was associated with the relative size of the hippocampus: experienced taxi drivers had a larger hippocampus than others.[42] In people suffering from Alzheimer's disease, by contrast, the hippocampus is found to shrink.[43] If only Proust's brain had been preserved as Einstein's was, it would have been possible to determine whether Proust also had an enlarged hippocampus.

Scientists who took Proust's descriptions seriously made a number of fascinating discoveries about the Proust effect in general and smell memories in particular. In conclusion, I will briefly recap and place them in perspective. Smell memories are found to arouse the strongest emotions of all sense memories, though they do not differ in terms of their dynamism and accuracy from image and sound memories, for example. A smell triggers a different, more intense and more emotional memory than reading the name of that smell on the label of a perfume bottle.

The Proust effect has proved to be virtually impossible to demonstrate in the laboratory because it occurs involuntarily and is not something that can be learned in advance. Autobiographical accounts are found to be a more suitable source for studying the dynamic of the Proust effect. For example, the Proust effect is found to occur gradually: first there is a sensory stimulus, then a vague sensation, which only after a period of time lasting from a few minutes to several hours is fleshed out with sensory experiences from the past, so that it can be localized in place and time.[44]

The Proust effect occurs largely outside the conscious mind. At the same time, there is a great deal of subconscious activity taking place in the brain, especially in the implicit memory which processes emotions and reports to the conscious mind only sporadically. This is why it is so difficult to control the Proust effect when it is happening: it appears to simply overcome the subject and is almost impossible to direct.

A Proust effect involves activity in many areas of the brain, partly because a great deal of information from the present and past comes together to construct the experience. Three crucial areas of the brain have a clear function here: the hippocampus collects memory pathways in different sensory domains; the amygdalae add colour and depth to the experience; and the frontal lobe oversees the coherence and consistency of the memories. These neurological processes can be compared with the creative process experienced by an artist. The Proust effect is not a faithful replication of the past, but a composition of sensory impressions, emotions, and physical feelings, which together give new meaning or insights to an episode from someone's past.

This was roughly how Proust thought of it a hundred years ago, and modern neuroscientists concede that he was broadly correct, based on recent insights into the neurological functioning of the memory. One of Proust's great achievements is that almost a hundred years after the publication of *In Search of Lost Time*, literary scholars, psychologists and neuroscientists still draw inspiration from and find challenges in his descriptions of the memory of the senses. The power of Proust's reflections is that he had an eye for the bigger picture.

Proust told only one side of the story: the calling up and retrieval of a past that had been lost. But how do children lay down their special experiences in their memory for later?

Sense memories more often go back to childhood memories in the period between 6 and 11 years of age. From the age of 11 onwards, memories are more often recovered through words, notions, and language associations. How are sense memories created in early childhood? The next two chapters provide perspectives on the production of sense memories in the first 10 years of a child's life. Before we explore neuropsychological studies on the subject, in Chapter 7 Vladimir Nabokov, one of the greatest storytellers of the twentieth century, talks about precisely this topic.

7

Nabokov as a toddler in St Petersburg

Stories of the origins of sense memories in childhood

'*In Search of Lost Time* is an evocation, not a description of the past' was how the Russian writer Vladimir Nabokov described Proust's magnum opus.[1] That comment is a common thread running through Nabokov's own autobiography, *Speak, Memory*, in which he describes how he explored his first childhood impressions and found true 'evocations' in the images, sounds, tactile impressions, and tastes of the early years of his life. His use of the term 'evocation' rather than 'description' shows that Nabokov did not take childhood memories as facts but rather as creative constructions of the mind. Fact and fiction are indistinguishably mingled at the very moment that memories come to mind.

What do those first impressions consist of precisely, and what remains of them when we remember them many years later? Do we store small photos of the past in our brains, or have we coloured in the black and white snaps from our parents' photo album and stored them in our brains in their new form? Nabokov's descriptions of his first childhood memories illustrate how the earliest impressions arise which are later combined with nostalgia, happiness, or other emotions to form sense memories.

Vladimir Nabokov was born in the last year of the nineteenth century into the family of a prominent Russian politician in St Petersburg. Two years after the Russian Revolution, the family emigrated to Western Europe, where Vladimir spent several years studying in Cambridge. Later he worked as a teacher and writer in Berlin, where he experienced the turbulent 1930s from close quarters. In 1938 he moved to Paris, and in 1940 he emigrated to

New England in the United States, where he spent much of his life working at Cornell and Harvard universities. He returned to Europe once and for all in 1958 and lived in the Swiss town of Montreux until his death in 1977.

His years in St Petersburg occupy a large number of pages in his autobiography, and contain a number of strikingly detailed descriptions from his first years of life—all the more remarkable since we know that adults' memories usually go back no further than the third year of life (figure 7.1).[2] Nabokov described his impressions as a child with photographic precision:

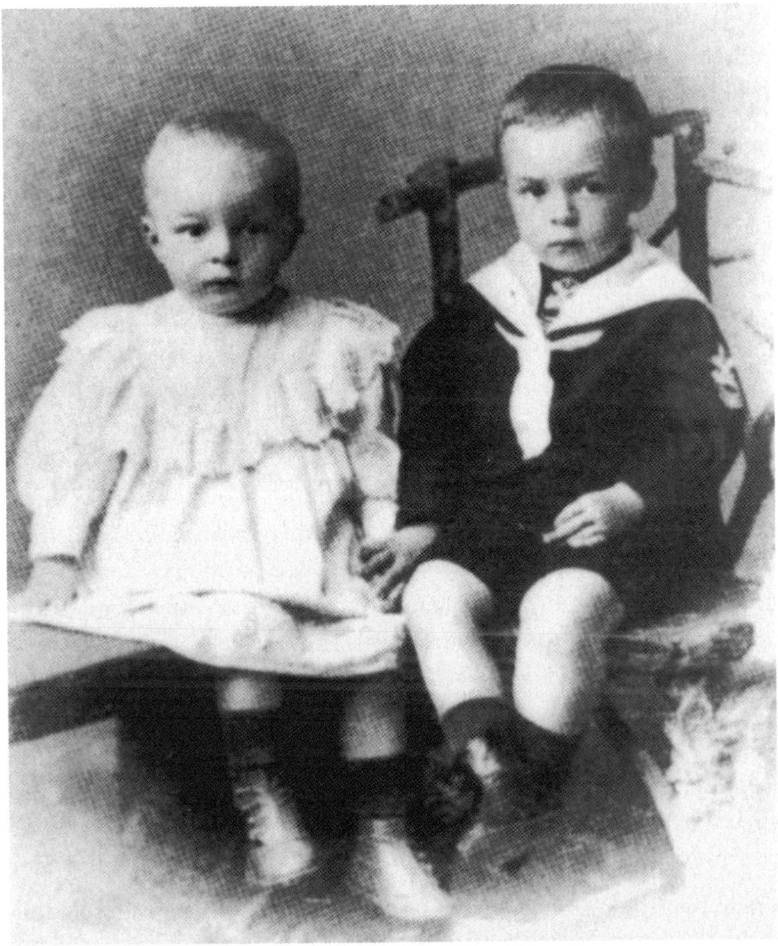

Figure 7.1. Vladimir Nabokov as a 2-year-old child (right) next to his younger brother Sergei (Biarritz 1901). Photograph taken from *Speak, Memory*. Copyright ©Vladimir Nabokov, 1947, 1948, 1949, 1950, 1951, 1967, used by permission of The Wylie Agency (UK) Limited.

> The recollection of my crib, with its lateral nets of fluffy cotton cords, brings back, too, the pleasure of handling a certain beautiful, delightfully solid, garnet-dark crystal egg left over from some unremembered Easter; I used to chew a corner of the bedsheet until it was thoroughly soaked and then wrap the egg in it tightly, so as to admire and re-lick the warm, ruddy glitter of the snugly enveloped facets that came seeping through with a miraculous completeness of glow and colour. But that was not yet the closest I got to feeding upon beauty.[3]

In this passage, Nabokov provides a detailed description of taste and tactile impressions which appear to be barely separated from each other in the writer's memory. With visible pleasure he describes the sensory impressions as an aesthetic experience, making no effort to separate fact from fiction. His memories were constructions, not descriptions. The same applies for the next fragment, when the young Vladimir was able to crawl—and thus to substantially expand his sensory world. Like every child, he loved constructing tunnels through which he could crawl in the living room:

> I then had the fantastic pleasure of creeping through that pitch-dark tunnel, where I lingered a little to listen to the singing in my ears—that lonesome vibration so familiar to small boys in dusty hiding places—and then, in a burst of delicious panic, on rapidly thudding hands and knees I would reach the tunnel's far end, push its cushion away, and be welcomed by a mesh of sunshine on the parquet under the canework of a Viennese chair and two gamesome flies settling by turns.[4]

We are again struck by the level of detail with which Nabokov conjures up memories before his mind's eye. With great pleasure, he re-immerses himself in the past. It is as if he is once again occupying that small child's body and just as easily feels the sound impressions in his ears as when he saw those two 'gamesome flies' before him. And he clearly enjoys the memory, which appears to be shot through with beauty and childlike emotion: 'a delicious panic'.

As soon as the younger Vladimir was able to walk, he would occasionally walk hand-in-hand with his mother and father, whom he remembered not as in a film, but as a mixture of colourful images and tactile impressions:

> Judging by the strong sunlight that, when I think of that revelation, immediately invades my memory with lobed sun flecks through overlapping patterns of greenery, the occasion may have been my mother's birthday, in late summer, in the country, [...] At that instant, I became acutely aware that the

> twenty-seven-year-old being, in soft white and pink, holding my left hand, was my mother, and that the thirty-three-year-old being, in hard white and gold, holding my right hand, was my father. Between them, as they evenly progressed, I strutted, and trotted, and strutted again, from sun fleck to sun fleck, along the middle of a path, which I easily identify today with an alley of ornamental oaklings in the park of our country estate, Vyra, in the former Province of St. Petersburg, Russia.[5]

The colours of the light appeared to mix with tactile impressions, so that it was almost possible to caress the softness of the colours, according to Nabokov's sense memories. The different sensory impressions worked in harness to produce a clear memory of his parental home.

Senses going back in time together

Like Proust, Nabokov relived a lost time in his memories. The descriptions of the two authors show similarities and differences. Both are sensory, emotional, and evoke feelings of pleasure and happiness. And like his French counterpart, Nabokov is grateful that he built up a childhood treasure-trove of aesthetic impressions in his childhood years:

> I may be inordinately fond of my earliest impressions, but then I have reason to be grateful to them. They led the way to a veritable Eden of visual and tactile sensations.[6]

There are also differences. Nabokov did not wait until the memories overcame him during lost moments, but went deliberately in search of the earliest impressions in his memory. Even more than Proust, Nabokov constructed a literary mix of fact and fiction, partly consciously and partly subconsciously. Where Proust devoted a great deal of attention in his descriptions to the physical and mental setting in which the memories arose, Nabokov's descriptions offer more insight into the evocations, the world of the evoked memories. For example, he described how sensory impressions melded together in his memory. Tactile impressions, in particular, always appeared to be present alongside Nabokov's various taste, sound, and colour memories. He was always aware in his memories of having a body that was reliving the past. That is something different from looking at a mental film of one's youth, adopting a bird's eye view of one's childhood, or looking at it as a third person might.[7]

Nabokov's descriptions showed how sensory impressions offered a means of accessing an emotional past, to which other sensory impressions were gradually added, transforming it into an aesthetic experience and giving rise to the pleasure with which those memories were recalled. He described the creation of the memory as a unifying of sensory impressions:

> In probing my childhood (which is the next best to probing one's eternity) I see the awakening of consciousness as a series of spaced flashes, with the intervals between them gradually diminishing until bright blocks of perception are formed, affording memory a slippery hold.[8]

This description of the melding of sensory impressions in Nabokov's childhood memories seems to refer to his synaesthesia. Synaesthesia is a neurological process involving cooperation between the senses in the brain, causing people to experience unusual sensations, such as seeing colours when hearing sounds. The term is a combination of the Ancient Greek words σύν (syn), 'together' and αἴσθησις (aisthēsis), 'sensation'.[9] Nabokov claimed that he also suffered from this 'medical aberration'. In *Speak, Memory*, he wrote:

> The confessions of a synesthete must sound tedious and pretentious to those who are protected from such leakings and drafts by more solid walls than mine are. To my mother, though, this all seemed quite normal. The matter came up, one day in my seventh year, as I was using a heap of old alphabet blocks to build a tower. I casually remarked to her that their colors were all wrong. We discovered then that some of her letters had the same tint as mine and that, besides, she was optically affected by musical notes. These evoked no chromatisms in me whatsoever.[10]

Principal characters in several of his novels have Nabokov's form of synaesthesia. In his novel *The Gift*, for example, he described the life of the young Russian émigré Fjodor, who displays striking parallels with the young Nabokov in Berlin (but who must not of course be confused with him, as the author warns us in his foreword). Fjodor recounts how since his youth he has been impeded by 'an extremely intense and refined *audition colorée*':

> For instance, the various numerous a's of the four languages which I speak differ for me in tinge, going from lacquered-black to splintery-gray like different sorts of wood. I recommend you my pink flannel m. I don't know if you remember the insulating cotton wool which was removed with the storm windows in spring? Well, that is my Russian y, or rather ugh, so grubby

and dull that words are ashamed to begin with it. If I had some paints today I would mix burnt-sienna and sepia for you so as to match the color of a gutta-percha ch sound.

Mozart in the Sistine Chapel

Synaesthetic processes cannot, however, explain why Vladimir Nabokov's memories were so detailed and intense. This can be understood better as a result of eidetic processes in his remembering, a second 'gift' from which Nabokov 'suffered'.[11]

Eidetic experiences can be described as a mix of current perceptions and memories and fantasies that are so intense and lifelike that they are perceived as if they were facts. The word 'eidetic' comes from the Greek word εἶδος (eidos), meaning image or form. Children at play, for example, frequently mix perceptions of objects such as dolls, chairs, and beds in the 'now' with memories and fantasies of people. The 'inanimate' objects come to life in their imagination as people, and this applies not only for dolls, but also for beds, chairs, and curtains, which can come to life as dragons, for example. Synaesthetic and eidetic experiences can be regarded as perceptual abilities that play an important role in the way children record experiences, as we shall see in Chapter 8.[12]

Both synaesthetic and eidetic experiences are often compared with hallucinations, because there is no observable and measurable object in the outside world that explains the sensation. They are not hallucinations, however, because accounts of synaesthetes and people with eidetic memories show that the sensations are in all cases a consistent and coherent part of their normal observational and experiential world.[13]

An appealing historical anecdote about which scholars still argue concerns the young Wolfgang Amadeus Mozart. The source is a letter from his father Leopold to his mother Anna, which describes a visit by Wolfgang and him in Rome in 1770.

An exceptionally beautiful version of the *Miserere* was sung in the Vatican each Easter in the Sistine Chapel. The *Miserere* had been composed a century earlier by Giorgio Allegri and was performed by two choirs of nine voices. The piece was written exclusively for the Sistine Chapel and was not permitted to be sung anywhere else or to be published in written form.

The young Wolfgang heard the *Miserere* with his father during a mass and later amazed everyone at their inn by writing out the whole piece of music at once. His proud father wrote to Wolfgang's mother Anna: 'Perhaps you have heard of the famous *Miserere* which may not be made public on penalty of excommunication. Well, we have it. Wolfgang wrote it down from memory'. The rumour spread rapidly around Rome that there was a child who could remember music after hearing it only once. The news also reached the Pope, and instead of ex-communicating Wolfgang, Pope Clement sent for him several times and showered him with medals and titles.[14]

Strikingly, Mozart appears to have heard sounds in colour and used the colours to remember music. There is no scientific proof for his synaesthesia, but it is known that Mozart wrote out a number of scores in four colours of ink (black, red, green, and blue). For a long time it was thought that Mozart did this just to be contrary, but closer study of the coloured scores shows clearly that there is a system in the use of colours.[15]

It remains a matter of speculation, but Mozart must have had an exceptionally detailed sound memory in order to be able to write out the *Miserere* in full after hearing it just once. Some form of synaesthesia will undoubtedly have been involved here.

Smelling and feeling memories

Medical literature contains descriptions of other, less famous examples of eidetic memory which provide a greater insight into the special features of these experiences. One early case dates from the 1930s. The psychologist D.M. Purdy from the University of California in Berkeley had a young female patient, whom he called R., who recounted a number of experiences which, though extraordinary, for her were normal and everyday.[16] She described being in hospital as a 10-year-old. The doctor asked her to imagine an image of the sun; she did so and had the feeling that she was staring directly into the orb. After a few moments it became painful and tears appeared in her eyes from the fierce sun. When she stopped imagining the sun, she could still see an after-image in the form of a purple disc before her eyes.

R. was able to recall images of trees and landscapes after studying them briefly and to describe those images with amazing accuracy. She described how the images appeared before her eyes on a translucent screen as if she was looking at them there and then. She could also look through that imaginary screen to the real world beyond. Strangely enough, she felt that the memories of the images were more lifelike and deeply coloured than the originals, and they also elicited other sensory impressions, as is the case in synaesthetic perceptions. When R. imagined an insect crawling on her arm, she could actually feel its legs and goose bumps appeared on her arm. When seeing a picture of a hunter pointing his rifle at a rabbit, she heard the rifle go off with a sharp crack and also saw and smelt the gunpowder smoke. If she thought of a rose, she could smell the scent just as clearly as she could see the deep red colour and feel the velvety petals. And finally, if she called up a picture of surf, she could smell the sea in her nostrils and hear the breaking of the waves on the beach.

Memory, imagination, and synaesthesia are closely linked, and were almost indistinguishable for R. Her physical reactions show that that she could feel a memory of an event almost as physically as if it were taking place there and then. As with Nabokov, it is impossible to separate fact from fiction, observation from imagination here. Memories are emotionally charged moments of sensory impressions relating to the recalled event. They are not faithful representations, because the imagination steers the construction of the memory. When remembering the picture of the hunter, R. was able to hear the rifle fire and, however invented that may be, the smell of gunpowder smoke was no less real in her perception and stimulated the inside of her nostrils.

More recently, the neurologist Richard Cytowic from Washington, DC described the case of a female scientist who underwent a major medical procedure.[17] The woman, designated as S. M., talked about her vivid memory of sounds, images, and smells. Her memories were so detailed that she was able to recall complex prescriptions by her doctor in such detail that it was as if she was listening to him talking. She could recall with ease every school classroom or living room she had been in in great detail, right down to the cracks in the wall and bits of loose wallpaper. She explained that she had an accurate memory for settings, colours, people, clothing, and so on. She was able to remember colours that she had not seen for many years. For example, when she wanted to buy a bead necklace for her mother who was visiting from abroad, she was able to find precisely the colour that matched a particular dress of her mother's; she was able to recall the colour of the dress in her memory in a precise and vivid way, even though she had last seen the dress 7 years previously. Similarly, 6 months after making a trip through Europe, she was still able to recall precisely the many hotel rooms she had stayed in, their furnishings and paintings. When she interviewed people as part of her job, she was able to recall the conversations in her memory 'to the last iota' and could replay them later in her mind as if they had been recorded.

After being diagnosed with a brain tumour, she was treated with radiation and consequently lost her sense of smell. This affected her ability to reconstruct the past in detail, which was based in some measure on her smell memories. However, she stated that her memories did not change during the treatment and that whenever she perceived an odour (even though she could no longer smell it), the images appeared again immediately. The difference now was that they no longer appeared unbidden, but that, like anyone else, she first had to 'think' about the smell in order to evoke the images. And her memories were also different; they too no longer appeared spontaneously, but she had to make an effort by consciously imagining the event. The content of her memories was the same, but they had become more faded, less vivid and intense than previously. This sad case shows that eidetic memories are dependent on the senses. In this woman's case, switching off one of the senses reduced the lifelike nature of the memories.

These accounts illustrate the creation of childhood memories as a result of synaesthetic and eidetic impressions. They mark a starting point for gaining an insight into how children's brains lay down notable experiences for later. In Chapter 8 we shall see which psychological and neurological processes play a role in this. Based on unusual and sometimes strange or amusing scientific experiments, we will make a journey deep into the human brain.

8

The little bricoleur

How children create eidetic and synaesthetic memories

Detailed, lifelike memories such as the eidetic memories discussed in Chapter 7 are often described as 'photographic' memories. This is not correct, because people with eidetic memories do not mechanically take 'snapshots' of light beams that strike their eyes and later retrieve those snapshots to view them again in the same way. Eidetic people can look around in their memories and describe details, but it is not like watching a film passing in front of their eyes. As an example, chess players who were said to have a photographic memory were in fact found to remember chess settings on the basis of the structures, not by conjuring up mental pictures of the settings. Sometimes they had not even seen the chess settings on the board, but had deduced them mentally.[1]

In the 1920s and 30s, the period when Nabokov was living in Berlin, psychologists led by Erich Jaensch at Marburg University discovered that eidetic memories in children are the result of a strong mix of daily perceptions with mental imaginings and memories.[2] From tests involving large groups of children, they found that children with strong eidetic abilities had a greater capacity for empathy. Based on his studies, Jaensch estimated that one in eight primary school children experienced intense eidetic sensations, and that the number of eidetic children in each age group declined rapidly during the secondary school period.

Jaensch's contemporary and colleague Heinz Werner placed eidetic and synaesthetic experiences at the centre of his theory on the development of thought in children.[3] He hypothesized that thinking and perceiving in children develop in accordance with the principle of 'differentiation and hierarchical integration'. For example, a toddler initially refers to all animals as 'doggy', but later also begins identifying cats and cows as well (differentiation), and even later begins to rank dogs, cats, and cows within the category 'animals' (hierarchical integration). Children's thought thus undergoes a sort of hop, skip, and jump process in which it becomes at the same time more subtle and more systematic.

According to Werner, synaesthetic and eidetic observations are undifferentiated impressions which are later differentiated and integrated into the thought processes of children—something for which physiological evidence has recently been found in neurological research.[4] However, Werner ignored adults who retain their synaesthetic and eidetic abilities beyond childhood. Recent research has shown that synaesthetic and eidetic experiences are integrated in the thought patterns of these individuals, who experience the phenomenon as perfectly normal.[5]

The magic broom

Eidetic and synaesthetic experiences are characterized by an empathic mode of observation in which the distinction between fact and fiction and between the internal and external world is fairly vague.[6] Children often experience their environment as a 'waking dream'; they draw a distinction between their own body and an object outside it less often than adults; as a result, objects are charged with emotions, because the child imbues the object with the movements and emotions of its own body.[7] The eidetic perception of a doll, for example, is more than simply a registration of the object, but is at the same time an emotional experience of the doll as a person. Think of the Disney films so popular with children, in which inanimate objects such as brooms and carpets come to life as people—e.g. the magic broom in *Fantasia* and the flying carpet in *Aladdin*. It is no coincidence that these films are so popular with children: they fit in perfectly with their experiential world. Children draw less of a distinction between reality and their own emotions; a broken stick, for example, can feel pain.

When I collected my old playpen from my mother for my own newly born son and saw the wooden bars, it immediately brought back memories of the time I spent in this child-friendly 'prison'. A small abacus was built into the front, with coloured wooden beads on horizontal metal rods, now bent with age and use. I was able to feel again how I pushed the beads back and forth endlessly with my baby hands (and never learned to count them), tasted once again the varnished wood (did no good, you chewed it off in an instant), and heard again the creaking of the hinges in the wood which allowed the pen to be folded flat. I realized how emotionally charged this old playpen was, which anyone else would have thrown out with the rubbish. The playpen simply came back to life and I felt once again precisely how it had helped me (piling up building blocks against one side) and also got in my way as I tried to reach something through the bars and got my arm stuck fast (stupid playpen). The playpen wasn't just an object; it was a person, I realized, a friend who was both outside me and within me.

Artists make use of empathic observations to give expression to the lines and colours in their compositions. The Dutch artist David Pierre Giottino Humbert de Superville, for example, developed a theory of forms in the early nineteenth century which was based on the emotions of lines.[8] A few decades later, the French painter Georges Seurat consciously used upward and downward lines to give his paintings a happy or sad expression. During the period of Abstract Expressionism in the twentieth century, artists such as Wassily Kandinsky and abstract filmmakers like Oskar Fischinger experimented with the empathic expression of lines, colours, and composition. And any number of such examples have been described by art historians.[9]

Werner gave an example of a personal memory of the artist Wassily Kandinsky, a contemporary of his, who was aware of his synaesthetic and eidetic abilities. He described how tubes of paint came to life for him as a teenager:

> On my palette sit high, round raindrops, puckishly flirting with each other, swaying and trembling. Unexpectedly they unite and suddenly become thin, sly threads which disappear in amongst the colors, and roguishly skip about and creep up the sleeves of my jacket. [...] As a thirteen- or fourteen-year-old boy I bought a box of oil-colors with pennies slowly I had painfully saved. To this very day I can still see these colors coming out of the tubes. One press of my fingers and jubilantly, festively, or grave and dreamy, or turned thoughtfully within themselves, the colors came forth. Or wild with sportiveness, with a deep sigh of liberation, with the deep tone of sorrow, with splendid strength and fortitude, with yielding softness and resignation, with stubborn self-mastery, with a delicate uncertainty of compassion, these curious, lovely things that are called colors.[10]

'Three is a jerk'

Scientists and others sometimes express doubts about the authenticity of eidetic and synaesthetic experiences, asking if these are not simply temporary children's fantasies. For this reason, Dan Smilek and his colleagues at the University of Waterloo, Canada studied a 17-year-old girl who saw numbers in an unusual way. Her account was as follows:

> Three is pure blue, the same color as E. Three is male; definitely male. Three is such a jerk! He only thinks of himself. He does not care about any other numbers or anything. All he wants is to better himself and he'll use any sneaky, underhanded means necessary. But he's also pretty young; he doesn't understand anything and he doesn't have very much power, as far as social status is concerned. So, he tries to hang out with Eight (who's also a bad number) just so that he can feel better about himself. But really, none of the numbers can stand him. He's a real jerk. He'll pretend as though he's your friend, but then he'll manipulate you and stab you in the back if he feels he can gain something from it. Then he'll never speak to you again. If Three had parents, even his parents would hate him. It's not as though what he does has some purpose or something behind it, he's just a really nasty number. He just wants things for himself. He doesn't care in what he does. If he had a voice, it wouldn't be high-pitched, but it wouldn't be deep. It'd be on the high side, a very annoying voice. He'd be short and very thin; very annoying.[11]

Few scientists would take this hilarious account seriously, but Smilek and his colleagues were intrigued by the fact that the girl saw numbers as personalities—and in fact not just numbers, but also letters and the furniture in her room. At her request, her parents had, for example, actually removed some items of furniture from her room because she felt they had extremely negative personalities. For example, she did not like the personalities of the number 3 or the letter S, while the number 2 and the letter M had pleasant personalities.

In the laboratory, the researchers subjected the girl to a series of psychological tests. For example, they showed her matrices containing numbers and letters—about which she commented in passing that it was like looking at people at a party—and noted how long she spent looking at the matrices while performing arithmetical tasks, and compared this with the times recorded for other people. They found that she looked at the matrices for longer than others and she recounted that some of them showed interesting groups of personalities. The time she spent looking at the matrices indicated that she had more difficulty detaching herself from symbols with negative personalities than from symbols with positive personalities.

The experiment also showed that she was not inventing the personifications of numbers on the spot. The number 3, for example, was and remained a 'jerk'. And when the researchers gave her new symbols and objects, her personifications remained consistent in the subsequent tests. Comparisons with other subjects who underwent the same tests showed that her scores always deviated very markedly from the average.

At around the same time, two researchers from the University of Edinburgh studied an individual with comparable empathic perceptions. Julia Simner and Emma Holenstein tested a 23-year-old female student who had different forms of synaesthesia.[12] Letters, numbers, and days of the week automatically generated very specific colour perceptions in her. She also experienced tactile impressions when tasting things, saw shapes in association with pain, and saw figures arranged in a visual pall of numbers.[13] The researchers in this study concentrated on her personifications of numbers and letters, which they termed *ordinal linguistic personification* (OLP). The female subject described the female (f) and male (m) numbers and letters as follows:

1 m: 'a good guy'; responsible; father figure; nice; a little tired.
4 m: good person; energetic; young; lots of 'get-up-and-go'; well-balanced.
5 f: mother figure; funny by accident; does things around the house.
7 m: submissive, weak man; not very confident.
8 f: fat; dating 9 but loves 7 (who'd be a great match but everyone prefers 6 with 7).
e m: 'a cheeky chappy'; talks when he doesn't know what he's talking about.
f m: 'a dodgy geezer' (i.e., untrustworthy); has connections; like e but less nice.
i m: little guy; h and g are always fussing over him; independent; wants to get on.
k f: energetic; bubbles along; not always approachable; mother of shy/quiet son l.
m f: old lady like n; they spend all their time together and gossip a lot.[14]

The psychological tests showed that her personifications were consistent (she never made a mistake in the personification of a number or letter) and that she reacted more slowly when a coloured letter appeared on the screen which did not correspond with her own colour for that number. Her test results were markedly higher (consistency) and lower (reaction speed) than those of 40 other volunteers who underwent the same tests. Finally, it transpired that the personifications had arisen in her childhood.

The third experiment shows how tangible an empathic experience can be. The British psychologists Michael Banissy and Jamie Ward reported a new form of synaesthesia in which those concerned feel as if they themselves are being physically touched when they see someone else being touched. The strange thing about these sensations is that they are mirrored: if the synaesthete sees a person facing them being touched on the left cheek, they feel as if they themselves are being touched on the right cheek. It is as if they are standing before a mirror and touching their own right cheek and seeing the mirror image of their hand appearing to touch their left cheek.[15]

A group of 'mirror synaesthetes' and a control group took part in the studies. The reaction times of the synaesthetes were found to be substantially longer when the perceived touching of another person's cheek did not correspond with the actual touching of their own cheek, and they made more errors here than the control group. The researchers were thus able to confirm the authenticity of these experiences.[16]

Subsequently, all participants completed a questionnaire to enable their degree of empathy to be measured, separated into cognitive empathy (seeing other people's viewpoint), emotional empathy (response to other people's emotions), and social skills. The 'mirror synaesthetes' scored higher on all empathy scales, but it was only in the domain of emotional empathy that the differences were statistically significant.[17]

The researchers believe there is a neurological explanation for this phenomenon in the form of hyperactivity in the system of mirror neurons in the brain, that is responsible for the ability to show empathy.[18] The discovery of these mirror neurons is generally regarded as a breakthrough in neurological research.[19] This research also offers new insights into synaesthetic, eidetic, and empathic experiences of children that form the basis of many of their sense memories.

Mirror neurons

A breakthrough in neurological research occurred in the 1990s in a laboratory in the Italian city of Parma, where Giacomo Rizzolatti discovered that certain neurons in the brains of macaque monkeys which normally controlled motor activity also became active when the monkey saw another primate performing the same motor action. The electrodes recorded the same activity in the premotor cortex when the monkey saw someone else eating as when the monkey itself was eating. The photograph of the researcher sticking out his tongue and holding a monkey that is imitating him, is a striking example of this (Figure 8.1).

The mirror neurons are located in the premotor cortex of the brain (see Appendix). Since these neural systems are active when observing actions in others and prompt the display of imitative behaviour, they are called mirror

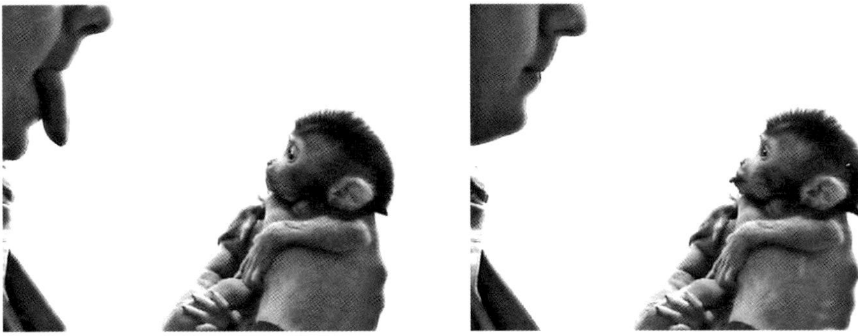

Figure 8.1. Newborn macaque monkey imitating human behaviour. (Source: Gross, L. (2006). Evolution of neonatal imitation. *PLoS Biology*, 4(9):e311. doi:10.1371/journal.pbio.0040311. Licensed under the Creative Commons Attribution-NonCommercial-NoDerivs 3.0 Unported License. To view a copy of this license, visit <http://creativecommons.org/licenses/by-nc-nd/3.0/> or send a letter to Creative Commons, 171 Second Street, Suite 300, San Francisco, California 94105, USA.)

neurons. This is a simplification, because in reality it is neural systems rather than individual neurons that are responsible for this.

It was not long after the experiments on monkeys that research on mirror neurons began in humans, too. Using a range of sensory stimuli such as seeing facial expressions, hearing creaking sounds and tasting unpleasant tastes, researchers went in search of mirror neurons in the human brain.

The research group at the Social Brain Lab at the University of Groningen in the Netherlands asked over 30 individuals to taste different tastes: pleasant (sugar), neutral (artificial saliva), and unpleasant (quinine). They were subsequently shown videos of the faces of actors who illustrated the tasting of the pleasant, neutral, and unpleasant tastes with normal facial expressions. Their brain activity was measured in both tests. The results showed that, apart from the taste centres in the brain, other centres were active which observe the person's bodily state, for example, feeling the heartbeat. A higher level of activity was measured in persons who scored highly on a psychological test measuring empathic ability.[20]

The mirror neurons also came into play when hearing sounds of actions, and again empathic individuals showed a stronger response. In another experiment by the same research group, they played different sounds to 16 subjects which referred to actions by other people: the sounds of kissing, gargling, crunching potato crisps, and tearing paper.[21] The subjects were then asked to carry out these actions themselves and the researchers compared their brain activity when hearing and when carrying out the activities. In both cases, the area of the brain that is sensitive to impressions from different senses was active. It was also activated when the actions were perceived visually.

In a third experiment, this research group measured not only the activity in the mirror neurons of 17 subjects as they perceived and imitated facial expressions, but also the activity in the amygdalae.[22] The more emotional the facial expression, the greater the level of activity in the amygdalae was found to be. This is not very surprising, because as we saw in Chapter 6, the amygdalae add emotional 'colour' to observations. This can explain the emotional empathy.[23]

Two other studies, one American and one British, offer a broader perspective on the activities that take place in the brain during empathic experiences. In both studies, the subjects watched a screen showing animations of abstract shapes such as triangles, squares, and lines. Two researchers from the University of Iowa studied a patient who had suffered damage to the amygdalae in both brain hemispheres. They asked her to look at animations of abstract figures and describe what she saw. Although tests had shown that she had normal linguistic and observational ability, she described the films in strict geometric terms, where others would recognize and describe the social interaction between the abstract figures. The amygdalae thus appear to play an essential role in social empathy, in the personification and anthropomorphization of abstract shapes.[24]

In the British study, researchers at University College London studied six subjects with intact amygdalae, recording their brain activity whilst they watched a silent visual animation of simple geometric figures.[25] Two figures appeared in the animations: a large red triangle and a small blue triangle, the movements of which against the white background could fairly readily be interpreted as social and emotional actions. The animations differed in the degree to which the geometric figures could be interpreted as 'human'. When seeing animations with a high potential for personification, four areas of the brain were found to be extra active: those responsible for attribution (cognitive empathy), the areas responsible for understanding the cartoons (and for things such as lip reading, for example), the areas responsible for processing emotions, and the areas responsible for observing movements.[26]

The research on mirror neurons began with macaque monkeys, and this immediately makes it clear that empathy is not the sole preserve of humans. In his book *The Age of Empathy*, the Dutch biologist Frans de Waal describes numerous examples of empathic behaviour in animals. One amusing example of this behaviour is that experienced by visitors to a dolphinarium, where dolphins respond to humans waving their hands by waving back with their flippers. Another amusing anecdote from De Waal's book describes a dog whose master temporarily had one leg in plaster; during this period, the dog also began to walk with a slight limp.

It was long thought that empathy was a quality that morally distinguished humans from animals, but of course nothing could be further from the truth. Humans and animals are essentially no different in terms of empathy and aggression. According to De Waal, the origin of the capacity for empathy lies in an older part of the brain known as the limbic system (see Appendix). This part of the brain controls emotional functions such as caring for others. The desire to care for others stems from the ability to empathize with another person's situation.

The founding father of research on empathy, the German psychologist Theodor Lipps, referred at the beginning of the twentieth century to what he called *Einfühlung*, which in later English interpretations of his ideas was translated as 'empathy'. This empathy, for example, feeling another person's pain or sadness, immediately awakens a feeling in humans and animals of 'I must help them'. This feeling is evolutionary, and is therefore very old and is one that we share with prehistoric animals. The mirror neurons are situated in a part of the brain which is more recent in evolutionary terms, the neocortex (outer layer of the brain), and De Waal regards them as a cognitive refinement of the more basic empathic emotions in the limbic system.

Control of empathic behaviour is found in different parts of the brain, suggesting that empathy is 'layered' rather than simply being a quality which a person either possesses or does not. De Waal distinguishes between different levels—and therefore functions—of empathy. At the personal level, empathy involves 'feeling for' other people or even objects, as we see, for example, with children who 'enchant' their toys by imbuing them with their own feelings and motives. At the social level, empathy is about the ability to put oneself in another's shoes, for example, feeling sadness during a funeral service, but also the almost irresistible urge to laugh when we see someone else laughing uncontrollably. Monkeys also do this, and the word 'aping' is the basis of what psychologists today refer to as 'mirror behaviour'.

At the societal level, empathy is the inclination to help others. In the arena of international politics, De Waal shows that the aggression involved in prosecuting wars is associated with an inability to empathize with others. A politician or soldier who is able to put themselves in the position of their opponent and empathize with him or her will have much greater difficulty in pushing the button or pulling the trigger than when their opponent is an unknown object.[27]

Children by nature have a rich empathic ability and, as adults, we naturally sometimes have doubts about the way in which children imbue their toys and friends with unrealistic qualities. Discouraging this display of empathy in young children seems to me a bad idea, however, which can have serious consequences for society because these children run the risk of developing into adults with little empathic ability. History is full of the most terrible examples of this.

A childhood treasure trove of memories

Children lay down emotional experiences for later in a different way from adults. Children more often use sensory empathic abilities (such as synaesthetic and eidetic experiences) to understand things—something that might be described as superficial imagination—whereas adults more often employ linguistic and logical techniques to fix experiences in their memories. This explains why childhood memories differ essentially from memories of our lives as adults. Where words and language offer easier access to adult members, sensory stimuli are more suited to tapping childhood memories. This may explain why sensory stimuli are able to trigger memories that go further back into early childhood than words and questions (see Chapter 6).

Synaesthetic and eidetic childhood memories are important for the development of empathy. They help children to empathize with the world around them and to understand and remember events.[28] Childhood observation is a process of constructing experiences in which sensory impressions are mixed together which adults normally try to keep separate. For example, the young Vladimir Nabokov draws little distinction between the objects around him and his own bodily feelings. A few clothes draped over a sofa become a tunnel which encloses his body. A crystal egg is not just a glittering image, but is also something that has taste, and those impressions in turn become mixed up with the wet-chewed sheet of his bed.

Children's imagination is part of their empathic development. Children build extraordinary treasure troves of memories during the first decade of their lives, on which they can draw with pleasure in later life. Children should be allowed to cherish such moments because later, when they are adults, they can prove to be important as a source of pleasure, as a basis for insight, and as an aid in processing emotions.

PART III

Practice

9

Do sense memories make you happier?

Personal well-being, aromatherapy, and taste lessons in school

What are the benefits of creating sense memories? Do they help us in daily life? Can we direct these involuntary childhood memories? Examples from psychological therapies, projects in primary schools (this chapter) and elderly care settings (Chapter 10) show how people can evoke and direct sense memories, with impressive results in increasing happiness and diminishing depression in young and older people.

Therapeutic smells

Aromatherapy is a form of alternative medicine whose origins lie mainly in and around France, the land of perfumes. In the town of Haasdonk near Antwerp in Belgian Flanders, smell specialist Veerle Waterschoot gives aromatherapy courses in which she helps participants investigate their personal smell memories based on smells which produce an unusual reaction in measurements of their skin tension.[1] It occurs quite regularly that one or two participants on her courses experience Proust effects.

I visited Veerle Waterschoot at her home in the Flemish countryside and, sitting in the garden, she told me about two experiences of participants which illustrate that Proust effects are not simply amusing, ephemeral sensations, but are capable of touching on existential issues for people. The first account was about a woman who suffered from anxiety and was not sleeping well. Examination showed a marked difference in her skin tension when she smelt Peru balsam than when she was exposed to other odours.[2] This prompted Waterschoot to begin working with her using this smell:

> I got her to smell the Peru balsam again, and once again she grimaced. She said: 'That smell is awful; it reminds me'—and her eyes filled with tears immediately—'of the cough syrup I had to take as a child because I was asthmatic. I had to take a great deal of that cough syrup. It made me sick.'
>
> In the afternoon she asked if she could take a break and go outside for a breath of fresh air. She looked very vulnerable, even though she was a strong woman, a manager in the business world. When she came back, she said: 'Veerle, I'm ready to work with that smell again. The experience of that horrible cough syrup is gone now, but I've rediscovered the loving care of my grandparents.'
>
> She told me that, as a child, she had spent a lot of time with her grandparents while her mother was away on business. 'Now I can feel why I found things so difficult with my mother, who was such a busy businesswoman, and to be honest I still blame her for that, but I realise that I've structured my own life in the same way and that my two darling children get very little from me.'[3]

According to Waterschoot, the smell unlocked an emotional problem for this woman, even before she was aware of it herself; it was only after she had been exposed to the smell that she was gradually able to become conscious of the emotional problem. Subsequently, she saw the background to the problem in the past and was able to link that to her present situation. The smell thus translated a vague anxiety into a tangible emotional problem, which she realized she had been carrying with her for a very long time.

That insight came 3 hours after smelling the Peru balsam. She recognized the smell immediately, even though Peru balsam is not common nowadays. Other participants, for example, were completely unfamiliar with the smell. All the woman was able to say at first was: 'I've come across that smell before'. At first, the moment of recognition only elicited a grimace of distaste. After 20 or 30 minutes, she realized that it was the smell of the cough syrup. And that was followed immediately by the account of her parents who ran a busy business. The story about her grandparents came 3 hours later, after she had asked if she could step out of the course for a little while to go for a walk. At that moment Waterschoot felt that she was vulnerable. When she came back from the walk, where she had probably felt great sadness, she talked about the love and care she had received from her grandparents and about her anger towards her mother. What she now understood was that she was essentially doing the same thing that her mother had done.

The second example given by Veerle Waterschoot was about an Italian man who was attending a 3-day course accompanied by his colleagues, all managers of a large telecommunications company:

> The purpose of the course was to strengthen the synergy and group cohesion. I gave out different smells and they were received with great hilarity. I then chose a fairly 'neutral' smell, a cedar wood fragrance, in order to calm things down. Suddenly one of the managers, an Italian by origin, became very emotional and said: 'This isn't possible; I've just been in my grandfather's wine cellar. I should think I was five years old, and I even felt his hand in my hand.' That immediately turned the atmosphere of hilarity into one of respect. He saw images and could smell the wine cellar. Later, in the bar, others said they had also experienced something similar, and just as intensely, but that they were too busy to acknowledge it. This man had allowed it in.[4]

In her practice, Waterschoot sees that when people experience a Proust effect, it often brings a highly charged emotion from the past to the surface. Proust himself also found the experience to be therapeutic, she believes:

> The Proust effect happens unexpectedly. You can't just sit down and make it happen. Two minutes can be enough to evoke an intense moment of being touched; that is inspiring and it keeps on coming back as long as the person concerned has not dealt with it. Our inner being recognises it as something important. At their young age, children aren't yet able to experience and understand it at such a deep 'soul' level, but they can still experience it as an emotion.
>
> From what I've seen in my practice, people start experiencing this type of Proust effect from around the age of 30. You have to be ripe for it. The recurrent nature of the effect is what gives it its therapeutic quality, allowing you to come to a state of harmony with yourself. A single experience is not enough for processing experiences in our subconscious; it's a lengthy process.[5]

Waterschoot describes the Proust effect as a moment of deep insight. She offers some of her own insight into the moment that past and present briefly coincide in a sense memory. She does not find it surprising that managers with a busy life are particularly prone to experiencing Proust effects during her courses: it is something that requires the subject to be in a calm, relaxed state. As we saw in Chapter 2, Proust experienced his sense memories at moments when he was calm, absent-minded, or bored. In moments of relaxation we are more aware of smells and other sensory impressions, and emotions from the past can also rise to the surface because they are no longer pushed away by current, acute activities.

Therefore, aromatherapists recommend naturally obtained essences based on their practical experience. They have a more calming effect on people than artificially produced smells, which therapists believe more often lead to aggression, irritation, and allergic reactions.[6]

Basic sensory training

The development of sense memories begins at an early age. As mentioned earlier, Proust pointed out that, if he had been less open to sensory impressions as a child, had not been able to learn to appreciate their beauty, he would not have been able to reap the rewards in later life in the form of the joy and insights that his memories brought him. This suggests that more could be done to foster aesthetic development in children, to help them increase their memory powers, as a sort of aesthetical capital for later in life. But how do you do that? Projects are being undertaken at a number of primary schools designed to enable children to discover and develop their aesthetical powers.

In Flanders, the government has taken a broad view of arts education which involves engaging all the senses in developing the tastes of primary school children. RASA, the Belgian national arts education foundation, commissioned the design of an unusual synaesthetic work of art to help children learn about the interaction of the senses.[7] Prik! (Stimulate!) is an installation built into a bright pink lorry which travels around the country (Figure 9.1). The children enter via a small staircase which admits them to a space where they step on to a conveyor belt covered in shag pile carpet that passes between moving, gossamer-covered walls through which various objects occasionally appear to be trying to protrude. In addition to these images, tactile textures, and movements, the children also encounter sound, music, and smells.

In order to build up an aesthetic childhood capital for later, children need to grow up surrounded by a wide variety of tastes and smells and to learn how to appreciate new and exotic tastes. If parents continue simply following the line of least resistance to the nearest fast food outlet and heating up meals in the microwave, there is little chance of this happening. All we remember of those meals later is that they were very sweet, very salty, and/or very fatty; in other words, nothing new and more of the same. It is precisely those first experiences with new tastes that stick in the memory. In response to this impoverishment of the sense of taste in children, some years ago the chef Pierre Wind, in collaboration with the Dutch Ministry of Education, Culture and Science, developed a series of taste lessons for primary schools.[8]

Figure 9.1. Exterior and interior of the Prik! trailer with a synaesthetic installation for children. Photos: Danny van Rumste, with permission from RASA, Belgium. See also plate 4.

Primary school children experiment with tastes, learn to become aware of them, and discover that other senses also play a role in how something tastes (Figure 9.2). These taste lessons have several purposes: they stimulate the senses, arouse children's curiosity to discover new things in relation to food, teach them to enjoy food, educate them about the origin and cultivation of the foods we eat and about the health aspects, and enable them to derive pleasure from preparing food. Or, as Wind summarizes it: 'The idea of taste lessons is not to hold up a finger and say 'this is how you must do it', but rather to allow children to get a broader picture of food and eating in a playful and experimental way'.[9]

Of course, this begs the question of whether the children will still remember these new impressions later in life. How might that work with smell memories? Occasionally one comes across a study that provides a hint. Researchers Jamie Ward and Julia Simner, for example, made a study of an eccentric Englishman named James Wannerton, who literally 'tastes' letters and words. He provides a very 'tasteful' account of his synaesthesia:[10]

Figure 9.2. Impression of a taste lesson for primary school pupils. Courtesy of CITAVERDE College, Herten-Roermond, The Netherlands.

> For as long as I can remember, words, word sounds, musical instruments and certain ambient noises have produced involuntary bursts of taste on my tongue.
>
> Texture and temperature also feature in this experience which is with me 24 hours a day. My dreams also contain tastes, and I am unable to turn it off.
>
> Although predominant during my formative years, I never considered these invasive sensations to be abnormal. Tasting words seemed as natural as breathing.
>
> As I got older and more involved in the wider world, I found my word/taste associations having an increasing effect in my everyday life, subtly dictating the nature and course of my friendships, personal relationships, my education, my career, where I live, what I wear, what I read, the make and colour of car that I drive. The list is endless.
>
> As a teenager, my developing interest in girls was heavily influenced by the tastes of their names, something which has led me up some disastrous paths over the years.
>
> A girl's name was just as important as her looks and in some extreme cases, even her personality.
>
> How shallow is that? I've constantly asked myself why girls with the sweetest tasting names quite often come with the sourest of temperaments.[11]

The researchers carried out a number of synaesthesia tests on Wannerton and discovered two remarkable characteristics of his synaesthesia.[12] First, they found that Wannerton more often experienced tastes with names of people he knew than with the names of strangers; and in general, words that elicited an emotional response in him more often had a taste than other words. Secondly, they discovered that the tastes associated with words were almost all tastes that he had first experienced during childhood. Words often tasted like breakfast products (such as cornflakes) and different kinds of sweets. He never experienced the tastes of coffee or curries, for example, even though he regularly drank or ate these as an adult.[13]

Like other children, James built up an aesthetic store house of taste associations in his childhood, even though in his case these associations were unusual because he linked tastes to words and letters through emotions. He is also unusual because he really does taste words, whereas in many adults the taste associations from childhood at most evoke a vague feeling, not an actual perception.

Some tastes are able to take us back to childhood and enable us to re-experience moments of sometimes intense emotion and lifelike sensory impressions of all kinds of images, sounds, tactile impressions, movements, and smells.

Thinking about a taste or its name is usually not enough to bring back memories in such intensity and sensory richness; it is usually essential to actually taste them. As the ingredients used in our food have changed and we consume more and more artificially manufactured meals, the likelihood of reliving tastes from the past appears to be diminishing. Many tastes from the past are in fact disappearing rapidly, and it was for this reason that the restaurateur and art historian Dirk Lambrechts called in his book *De smaak van heimwee: Herinneringen aan de Europese volkskeuken* ('The taste of nostalgia: memories of traditional European cooking') for tastes to be preserved in the same way as we are accustomed to doing with paintings and music that are considered to be of value.[14] Lambrechts has now passed away, but his idea has been institutionalized in the international Slow Food movement.

The movement began in Italy, where a university of taste has since been established in the hamlet of Pollenzo in Piedmont. Slow Food started as a reaction to the growth of fast food chains which were increasingly shaping the tastes of children, in particular. It was also a reaction to the constant imposition of ever more regulations by the European Union which, under the pretext of standardization and food hygiene, was decimating the range of permitted food products. Among the hardest hit were local products such as handmade cheeses, which derived their distinctive taste from local moulds. The Slow Food movement, which now has representatives in many countries, advocates the preservation of traditional tastes from the past.

In his book, Lambrechts travels through Europe telling tales of tastes that appear to be being lost and attempts to preserve them, at least on paper: 'Today, many years later, I feel like something of a curator of a past that has been irrevocably lost. A curator without a collection or archive, only of memories'.[15] Lambrechts presented some nice examples of taste memories from his own Flemish childhood:

> My first memory of food is the smell of bacon, fatty bacon of a kind that we don't get today. On Saturday evening a big pan of bacon was placed in one of the ovens of the stove, and the bacon spent all night cooking slowly. On Sunday morning fresh-baked rye bread was placed on the table (white wheat bread was eaten only on special occasions), along with the pan filled with what was by now almost translucent bacon, glistening in the gently bubbling bacon fat. Eaten with strong, black coffee, it made for a festive breakfast. People would call it a poor man's breakfast today, and it would give dieticians the shivers. But you have to remember that it wasn't an everyday breakfast and that in those days most people did heavy physical work.[16]

I still regularly experience this for myself. Decades ago, I visited traditional, what are now Slow Food restaurants in Piedmont with some Italian friends. Since then, whenever I experience the taste of truffle, which I first encountered there, or a pasta with a specific texture—the taste never works on its own—I am simply able to recall images, sounds, smells, and tastes of a small restaurant in Barolo, Piedmont, on a cold, dank, misty November day. The taste is strongly associated with that moment and the emotions of that day. It is distinct from other tastes and I probably really learned to taste for the first time there. I have eaten pasta often, and it creates memory pathways with many recollections—perhaps too many different memory pathways for my brain to choose from. By contrast, the taste of supermarket pasta evokes no memories for me, but has become abstracted from a palpable situation with images, sounds, tastes, smells, tactile impressions, etc., to nothing more than an isolated, abstract entity. The packaging of supermarket pasta merely reminds me of walking down the aisle with a supermarket trolley past the plastic bags of dried pasta.

Another taste memory—which I appear to share with Dutch people of my generation—was one I discovered when looking at the Memory of the Netherlands website.[17] The taste of *Brinta* breakfast cereal takes me straight back to my childhood. *Brinta* doesn't have just one taste, but several in my memory of the senses. As I wrote in Chapter 1, it depended on how successfully the 'porridge' was made. Too much cereal and too little milk produced a sort of cement; using milk that was too hot resulted in a slimy mess. But when the proportions were just right, the result was a delicious, creamy porridge with a slightly firm surface on which you could sprinkle hundreds and thousands and scoop them off again with your spoon.

The Memory of the Netherlands website consists mainly of images and sound material, but could just as easily incorporate the preservation of taste memories.

Children often find new experiences exciting and a little scary. This even applies when tasting new things; that is an experience shared by every parent and anyone who has helped bring up small children. By turning it into a pleasurable activity, children are able to enjoy new sensory impressions in an atmosphere of calm and trust. This is an argument for more intensive aesthetical education at school. Emotional and social success in later life demands a recognition of the importance of sensory education alongside the teaching of arithmetic and languages.

Children differ in their aptitude and ability to use their senses. Some children have more visual thought processes than others.[18] Some perceive the world more synaesthetically than others.[19] When teaching these children, it is important to make allowance for individual differences in their sensory abilities. Many tests are available at schools for measuring individual differences in cognitive skills (arithmetic, language), but there are none for measuring differences in sensory skills. At best, the teacher will notice that one child is able to draw a human being more accurately than another, or is able to learn to play a recorder more quickly than others. There is a challenge here for designers of teaching aids and tests.

10

Uplifting musical memories

People with depression, dementia, and care for older people

The New York-based British neurologist Oliver Sacks once remarked about musical memories in an interview: 'A large part of the effect of music takes place outside our conscious mind. It creeps inside, just like Proust's madeleines'.[1]

Which deeper layers of consciousness is Sacks referring to? In his book *Musicophilia* he presents a large number of anecdotes which provide a hint as to an answer.[2] Among the most intriguing subconscious pathways of musical experiences and memories are dreams. Sacks describes how composers such as Wagner and Berlioz dreamt music before they composed it and wrote it down.[3] And dreams need not produce music alone; it is not uncommon for people interviewed by researchers to recount that the music they have dreamt is also accompanied by images.

Why do melodies in dreams also elicit images? This would appear to be a form of synaesthesia.[4] The American visual artist and synaesthete Marcia Smilack, whom I visited at Martha's Vineyard, an island off the coast of Boston which is much loved as a home and holiday venue by writers, musicians, artists, and Democrat presidents, told me that she sometimes dreamt about musical colours:

> I also dream music often. Once, a favorite dream was the marriage of the browns. It's hard to explain but these colors of brown were getting married and as they walked down the aisle, I heard the most beautiful music I have ever heard in my life.[5]

Her daily work, too, is often based on the music of images: for example, she photographs reflections in the water when they evoke music in her head.

Plate 1 'Smelloflowers' in the Sensire Den Ooiman care centre in Doetinchem (the Netherlands) by artist Peter De Cupere, <http://www.peterdecupere.net/2013> (with permission from Peter De Cupere). © Peter De Cuper, 2013. See also Figure 3.1.

Plate 2 Wartime food by eating designer Marije Vogelzang, <http:www.marijevogelzang.nl> (with permission from Marije Vogelzang). © Marije Vogelzang, 2013. See also Figure 3.12

Plate 3. Left: *The Visitation*, 1528–1529, by Jacopo Pontormo (Church of San Michele in Carmignano, near Florence). Right: a still from the video *The Greeting* (1995) by Bill Viola (De Pont Museum, Tilburg). Video/sound installation, 430 × 660 × 780 cm. Colour video projection on large vertical screen mounted on wall in darkened space; amplified stereo sound. Performers: Angela Black, Suzanne Peters, Bonnie Snyder. Production still. Photo: Kira Perov. See also Figure 5.3.

Plate 4. Exterior and interior of the Prik! trailer with a synaesthetic installation for children. Photos: Danny van Rumste, with permission from RASA, Belgium. See also Figure 9.1.

Plate 5. Marcia Smilack, *Yellow Boat Minor Chord.* With permission from Marcia Smilack: <http://www.marciasmilack.com>. See also Figure 10.1.

Plate 6. Multisensory 'Beach room' in the Vreugdehof care centre, Amsterdam. Photo: Cor Mantel, with permission from Vreugdehof. See also Figure 10.2.

Plate 7. Painting of the first 20 digits of pi by Daniel Tammet. With permission from Daniel Tammet: <http://www.optimnem.co.uk>. See also Figure 11.1.

Plate 8. Janneke van Dijk, February 1953. *When the waters came…* She writes about the work:
'It is a chaotic image, as if the ground is falling away beneath me. The blue above is the colour of my father and the brownish tint below is the colour of my mother'. © Janneke van Dijk. See also Figure 12.1.

Plate 9. Clara Froger, *Not sleeping on the northern Rotterdam ring road*, 2009. She explains: 'Colour images that pass by in constant motion. Lorries and cars, trains in an unending stream of sound and colour and the alarm clock as a dotted line'. © Clara Froger, 2009. See also Figure 12.3.

Plate 10. Martina Strusny, *Viel mehr Rot als Blau* ('Much more red than blue'). Photo: Ron Zijlstra. © Martina Strusny, 2013. See also Figure 12.4.

She describes this phenomenon herself using the photograph entitled *Yellow Boat Minor Chord* (Figure 10.1):

> I took this image when I heard a yellow chord in a minor key. To my surprise, it appeared a few years later in one of my dreams. I was watching a sunset with friends when I noticed, mixed in with the regular sailboats on the other side of the pond, this reflection. There it was, a two-dimensional rendering, afloat, pretending to be the real thing. No one noticed but me and when I next looked at its shimmering it winked back at me.[6]

The musical synaesthesia that she experiences in her daily life has an important counterpart in her dreams. She often dreams the musical images before she photographs them:

> I dream many images before I actually see them on the water. I often feel like a reporter from the Collective Unconscious. I stand poised at the edge of a floating tableau of reflections that pass by whether anyone notices them or not. Collected, these archetypal images form a universal alphabet. Tide to tide, the sea's mirrored surface washes up to and over the sea-land barrier like a cerebral moat, as if the universe is dreaming. And while reflected images, like dreams, cannot be repeated, unlike dreams they can be recorded. I go fishing for archetypes and the catch that I bring home—the dreams of the universe—belongs to us all.[7]

Figure 10.1. Marcia Smilack, *Yellow Boat Minor Chord*. With permission from Marcia Smilack: <http://www.marciasmilack.com>. See also plate 5.

Her 'waking dreams', as she calls them, manifest themselves not only in synaesthetic perceptions of coloured music, but according to Smilack are a more general phenomenon of our consciousness:

> I just think that for me the walls within my mind are extremely permeable so there is a lot of exchanging of material between waking and sleeping, dreaming and consciousness. I honestly do not know where synesthesia fits into all of it but I do know that I have always been I am unusual in how much I dream; and so naturally my synesthesia shows up in my dreams (how could it not?).[8]

The American singer-songwriter Billy Joel was for a short time her neighbour at Martha's Vineyard, and when he came to look at her photographs one day, they fell to talking about the colours of music. It transpired that Joel also experiences music in colours and uses this as a means of remembering melodies that arise in his dreams. He told the American author Maureen Seaberg, who wrote a book about the synaesthesia of famous musicians and other creative individuals, that lots of ideas for songs appear in his dreams.[9] The problem is that he finds it difficult to recall the melody once he has woken up. He has occasionally tried sleeping with a tape recorder on his bedside table, but that didn't work. Then one day it occurred to him that the colours of the music could help him. Now, when he wakes up with the feeling that he has dreamt a melody, he first tries to recall the colour, and via the colour the melody gradually reappears in his mind:

> Mr. Joel says that sometimes songs will reappear the next day and sometimes they won't return for months or even years. 'The song "Just the Way You Are" is a song I had dreamt and forgotten." [...] The storyteller/bard goes on to explain that he wrote the song fairly quickly, then said, 'How the hell did that happen?' On reflection, he realized he was thinking he'd heard it before. Then the aha feeling—yes, it was a synaesthetic vision in a dream. The blues and greens and their various amorphous shapes had encoded it in his mind somewhere to be retrieved later.[10]

Music therapy

Smilack recounted how another friend who also lived on the island, the late American author William Styron, perhaps best known for his novel *Sophie's Choice*, often talked about dreams. Styron told her that a dream about Sophie had been the inspiration for the later novel, which deals with a depressed and troubled Polish woman in New York who is tormented by terrible memories of the concentration camps in the Second World War.

Dreams and the subconscious mind play an important role in the work of Styron, who was himself affected by severe depression. His book *Darkness Visible: A Memoir of Madness* presents a gripping account of his depression as a journey into hell.[11]

Sound memories can sometimes bring light into the lives of people suffering from depression who are sinking further into their own personal darkness. Styron is one of the cases cited by Sacks in *Musicophilia*. In the following quote, Styron describes the wondrous effect of the sounds of film music in the downward spiral of dark feelings during his depression:

> My wife had gone to bed, and I had forced myself to watch the tape of a movie. [...] At one point in the film, which was set in late-nineteenth-century Boston, the characters moved down the hallway of a music conservatory, beyond the walls of which, from unseen musicians, came a contralto voice, a sudden soaring passage from the Brahms Alto Rhapsody. This sound, which like all music—indeed, like all pleasure—I had been numbly unresponsive to for months, pierced my heart like a dagger, and in a flood of swift recollection I thought of all the joys the house had known: the children who had rushed through its rooms, the festivals, the love and work.[12]

Through the music, Styron suddenly and powerfully recalled the joyous moments of home, enabling him to leave behind the numbness of his depression. Smilack told me more about the personal background to this:

> Music was enormously important to Bill and I remember a line from one of his short stories in his last book in the final story where he tells of the death of his mother when he was a young boy; in it, he describes his mother's voice as an opalescent sound the young boy in the story could almost see, which reminded me of how I had often thought he was very close to synesthesia; but he insisted that for him, it remained a metaphor and that he never actually saw it. So that is why I thought he came close. Incidentally, his mother was trained as an opera singer and Bill used music all his life for inspiration for his writing. In fact, when he was depressed, one of the major symptoms was that he could no longer enjoy music or respond to it. Once, after a blood transfusion, he started to heal and suddenly asked Rose and me why there was no music available to him in the hospital room. Rose and I were both shocked because he had not wanted to listen to music for many months, so his sudden desire for it again reinforced for us the fact that he had needed the blood transfusion (he was quite anaemic) to treat the depression which allowed the pleasure from music to return to him.[13]

This account illustrates that in Sacks' deeper layers of consciousness, sound memories sometimes unlock emotions which seemed to have been locked away for ever. In another anecdote, Sacks talks about singing as a 'can-opener' for accessing memories in older people with dementia. An Australian music therapist in a nursing home wrote to him saying that she had initially started singing sessions as a means of providing some diversion for residents with dementia, and had discovered what effect it had upon them and how much joy it brought them:

> At first I thought I was providing entertainment, but now I know that what I do is act as a can-opener for people's memories. I can't predict what will be the trigger for each person, but there is usually something for everyone, and I have a part of my brain that 'watches' in stunned amazement what is happening....One of the loveliest outcomes of my work is that nursing staff can suddenly see their charges in a whole new light, as people who have had a past, and not only a past but a past with joy and delight in it.
>
> There are listeners who come and stand beside or in front of me, touching me, for the whole time. There are always people who cry. There are people who dance, and people who join in—for operetta or for Sinatra songs (and Lieder, in German!). There are disturbed people who become calm, and silent people who give voice, frozen people who beat time. There are people who don't know where they are, but who recognize me immediately as 'the Singing Lady.'[14]

The Music and Memory organization in the United States trains nursing home staff how to challenge people with Alzheimer's and dementia with music.[15] The Music for Life project in England invites older people to take up a musical instrument or join singing sessions with other older persons. The effects of practising music on well-being and participation in society have also been under study for some time.[16]

Sense and reminiscence in older people

Internal projects have been developed for various reasons in nursing and residential care homes which have been found to stimulate the memory of multiple senses, not just musical memories. For older people with longer memories, and with lots of both positive and negative recollections, the playful form of these projects has been found to strike a good balance between good and bad memories. We saw in Chapter 3 how older people took great pleasure in participating in the workshops on wartime food organized by

Marije Vogelzang in Rotterdam, while older residents with dementia in the *Den Ooiman* nursing home in Doetinchem in the east of the Netherlands are helped to find their way in their lives with *Smelloflowers*.

The 'memory museums' in the nursing and care homes operated by the Dutch Humanitas foundation were established to give the residents something to talk about (instead of complaining, according to the director), so that they continue actively using their memories and also remain socially active. The memory museum is full of objects, illustrations, smells, sounds, and documentation from the last century. For example, there is orange furniture from the 1970s, bicycle tax plates and distribution cards from the Second World War, but also petrol distribution coupons from the oil crisis, an old Solex power-assisted bicycle and a Puch scooter.[17]

Another means of stimulating the sense memories of older people is the 'Storytable', a piece of furniture designed by the visual artist Hans Muller around which about six people can sit. Using two knobs, short image and sound fragments can be played back. The material comprises songs, newsreel clips, and short films of day-to-day life. It is a sort of museum for sound and image in the living room. Residents, children, and activity supervisors can also add their own film material. The fragments stimulate sense memories, which in turn lead to discussions around the table. Here again, the aim of the homes is to reduce loneliness and bring more happiness into the lives of residents.[18]

A digital variant of the Storytable is the M-book, an interactive photo album which takes advantage of the fact that the ability of older people with dementia to 'recognize' often remains intact for a long time, by enabling them to look at old photos together and listen to familiar voices.[19]

These and other initiatives which stimulate sense memories can help even the most vulnerable older people. They provide fuel for conversation, promote social contact, and reduce loneliness. Perhaps no less important, however, is that they also boost older people's sense of self-worth, by reminding them that they still know things about the past of which young people have no inkling, can still remember what old objects were used for, etc. Older people who have been written off by society then suddenly appear to be more skilled and to know more than the younger generations.

Using simple objects which evoke sense memories enables older people to pluck the fruits of their aesthetic childhood memories, and this has been found to increase their pleasure in and happiness with life.

Happiness and depression

Does the memory of the senses genuinely make people happy? Proust was fascinated by the joyful feelings that sense memories evoked in him. However, sense memories are not by definition guaranteed to bring happiness; they can sometimes awaken feelings of anxiety, depression, and loneliness. They are almost always intense, emotional, empathetic, and lifelike, in both a positive and negative sense.

Sense memories have had a therapeutic effect for several people described in this book—not in the medical sense that they have cured people of their organic mental illnesses, but more in the sense of helping with the milder psychological problems which occasionally cross people's path as they journey through life. Everyone at some point in their lives encounters one or more moments of existential difficulty. Closer analysis reveals that they are almost always connected with an emotional past, and impediments that hinder their optimum functioning in the present.

Unfocused or simply unpleasant memories and events can crop up completely out of the blue, seemingly without any direct cause. Sometimes they can be temporarily suppressed by seeking diversion. But if they continue to recur, it will sooner or later be necessary to pay attention to them.

The active, positive use of sense memories as a means of promoting happiness received scientific support in a series of surprising experiments. Tali Sharot and her colleagues at New York University studied brain scans of optimists and pessimists and found that optimists make more use of their autobiographical memory than pessimists. In formulating pictures of the future, people use the parts of the brain that integrate emotions and autobiographical information. Those parts of the brain are more active in optimists than pessimists. By contrast, brain scans of people with depression showed irregularities in these parts of the brain.[20] In another study it was found that people with high self-esteem recall more details in memories. They reactivate positive memories more often than people with low self-esteem.[21]

The psychologist Fred Bryant and his colleagues at Loyola University Chicago studied the effect of memories on happiness in students.[22] In the first study, Bryant asked students to complete a questionnaire about pleasant experiences: the ability to enjoy things, how often they spent time recalling memories, what kind of memories they were, and when they did that. The results showed that the more often students recalled pleasant memories, the more able they were to enjoy life; the more sensory triggers they described for intense, spontaneous reminiscences, the more they had the feeling of enjoying life. Students who 'sought refuge' in memories as a means of avoiding present-day problems (a form of escapism) were less happy.[23]

To address the possibility that the questionnaire-based survey might be distorted by the way in which students viewed the subject of happy memories, Bryant et al. performed a second study, an experiment involving two groups of students.[24] The students were tested before the experiment to assess their degree of happiness and ability to enjoy life. They were then asked to sit alone in a quiet room twice a day for a week and to think about their lives in accordance with a list of instructions they had been given. One group received a list of memorabilia (souvenirs, photos, presents) and were asked to write down a positive memory for each object. The control group were asked to relax and think about their current lives. At the end of the week all participants were retested to assess their happiness and ability to enjoy life. The researchers found that the happiness of the members of the experimental group had increased while that of the control group remained unchanged. The biggest increase in happiness occurred in participants with the most lifelike memories. Their memories were more real, with lots of accompanying sensory impressions such as images, sounds, and smells. The realism of the memories also had a greater impact on their feelings of happiness than the number of details they were able to describe from the memory.[25]

If sense memories make people happier, might this be of help in combating depression? Depression is a condition that is more prevalent among older people. Older people suffering from depression have difficulty remembering specific events from the past. It is as if a dark blanket or haze of negative feelings overlays the past. In another initiative, using a simple therapy for residents of a care home, namely encouraging them to recall detailed memories of pleasant events from their lives, was found to reduce their depressive symptoms. The participants were given written instructions and asked to recall a pleasant situation from their youth. By being questioned about details, they were prompted to flesh out their memories to make them more tangible. The therapy was found to be more effective in people with severe depression than those with mild depressive complaints.[26] The authors give a touching example of one such memory by an 85-year old woman, living in a residential home in the centre of Amsterdam:

> When I was 14 years old we spent Christmas at the Salvation Army. We were very poor. We had a very good dinner of beans and bacon. It tasted really delicious.

> I will never forget the taste, smell, and sight of that meal. I went there with my father and my two sisters and three brothers. We sang songs during the Christmas celebration at the Salvation Army. They had a beautiful Christmas tree. They gave mandarins wrapped in red paper that we hung in the tree. I will never forget that Christmas. We were poor at home, my mother had passed away, we were hungry. We were very grateful for receiving the meal of beans and bacon.[27]

This is a typical example of a sense memory. There is a sensory stimulus—the taste or smell of beans and bacon—and an emotional carrier, the feeling of gratitude. The anecdote shows that the experiencing of sense memories can protect older people against feelings of depression.

A number studies have shown that involuntary autobiographical memories are generally emotionally positive and involve more mood impact and physical reaction at the time of recall than voluntary memories.[28] Other studies have shown that these kinds of involuntary sense memories, in particular, elicit specific events on a specific day in the past more frequently than voluntary verbal-cued memories.[29]

There is also a counselling method in which sensory stimuli are used. One category of these stimuli is 'smells from the past'. In 12 group sessions lasting between 1½ and 2 hours, under the name 'Looking for meaning', older persons were presented with sensory stimuli such as smells from the past and other materials (old sayings, reminders of homes where they had lived in the past) as a means of evoking memories. The purpose of the course was to bring out positive personal memories and allow the participants to re-experience the emotions associated with those events (reminiscence) and also to evaluate those emotions, process them, and obtain a personal insight into them. The participants also made collages together using the materials, thereby unlocking not only personal memories but also social conversation.

Did the project produce results? In a scientific study, the participants in this course were compared with a comparable group of older persons who had watched a film about learning to grow old successfully, entitled *The Art of Growing Older*. After the course, the participants had fewer depressive complaints and felt more in control of their lives than the control group who had watched the film.[30]

The course 'Looking for meaning' is a combination of two methods: reminiscence therapy and life review. In reminiscence therapies, older people come together to discuss events and activities from their personal past, aided by concrete objects such as photos, sound recordings, tastes, and smells. These objects stimulate the senses and memories of the participants and are intended to foster their well-being and self-awareness. Life review therapy involves a session in which a person receives guidance in re-experiencing and evaluating important events from their life. The participant then uses written accounts and collages of images to produce a 'life book'.[31]

Family members are often actively involved in reminiscence therapies. This is one of the most popular psychosocial interventions for people with dementia. An interesting sensory variant is *snoezelen*, about which more later. Participants and care providers are generally very enthusiastic about this form of therapy, and scientific research has revealed its positive effects on the well-being of care and nursing home residents. The effects on the well-being of people with dementia is also being studied, but because their well-being is so difficult to measure (or ask them about), there is no consensus about its effects.[32]

Reminiscence is the voluntary or involuntary recalling of personal memories. They have different functions, such as helping people to define their identity, solve problems whose origins lie in the past, or prepare for death (terminally ill patients). These functions often occur together and combine to form 'self-positive reminiscences'. There are also a number of functions that fall into the category of negative reminiscences (seeking refuge in the past, escapism): alleviating boredom, reliving bitter experiences (recalling unjust events to feed present feelings of bitterness), sustaining intimacy (trying to hold on to a loved one who has passed away). And finally, there are social functions such as educating and informing (transferring memories and life lessons to others) and conversation (engaging in contact with others by telling stories). These functions have different effects on well-being. The social functions contribute to people's happiness and sense of well-being; the identity-forming, self-positive reminiscences contribute to mental health, while escapist reminiscences are associated with depressive complaints.

The Storytables and the memory museums referred to earlier are examples of these functions. In addition to these 'free' forms of recollection, there are also slightly more coercive forms such as courses and even more coercive forms such as therapies. The more intense the form, the greater the effects found in reducing depression. Life review is now ranked alongside cognitive behavioural therapy as one of the two most effective means of treating depression in older people.[33]

The *Snoezelen* approach was initiated in the Netherlands in the late 1970s. The word *Snoezelen* is a combination of two Dutch words: *doezelen* (to doze) and *snuffelen* (to sniff). Consequently, the approach implies both a restful activity and a more dynamic sensory aspect. The sensory stimuli are offered to each client through visual, auditory, tactile, and olfactory modalities, and according to his or her personal choice. Snoezelen takes place in a specially equipped room where the nature, quantity, arrangement, and intensity of stimulation are controlled. The aim of the intervention is to find a balance between relaxation and activity within the framework of a safe, adapted environment, supported by a facilitator.[34]

Snoezelen has become very popular in nursing homes: around 75% of homes in the Netherlands, for example, have a room set aside for snoezelen activities. Two kinds of changes in residents are commonly cited. First, carers have found it easier to make contact with residents, in turn eliciting a greater response from the residents. Secondly, carers have found that residents who display challenging behaviour, such as agitation, restlessness, or aggression, become calmer and more settled; these are significant differences.[35]

The Vreugdehof residential care centre in Amsterdam has built a 'beach room' especially for residents with dementia (Figure 10.2). In this room, residents can enjoy the feeling of sitting in the sun with their bare feet in the sand. The room is designed to improve the well-being of these residents.

Figure 10.2. Multisensory 'Beach room' in the Vreugdehof care centre, Amsterdam. Photo: Cor Mantel, with permission from Vreugdehof. See also plate 6.

The garden room at the centre has recently been converted into a true 'beach room', complete with sandy beach and a 'sun' which can be adjusted in intensity and heat output. A summer breeze blows occasionally and the sounds of waves and seagulls can be heard. The décor on the walls is several metres high, giving those in the room the impression that they are looking out over the sea. There are five or six chairs in the room where the older residents can sit. There are also areas of wooden decking on which wheelchairs can be parked. The designers have even added sea air.[36]

One male resident used to go to the beach often in the past and now, after initially protesting when his daughter collected him from his bedroom, feels calm and content in the beach room. His dementia means that we cannot ask him whether he remembers anything from the past, but there does appear to be a moment of recognition of a familiar setting from the past.[37]

After these accounts of how sense memories contribute to frail people's personal well-being, Chapter 11 will show how sense memories can also help us to recall practical things that we want to remember, such as appointments, lists, and numbers.

11

Remembering 20,000 digits of pi

How memory artists use sense memories

The exceptional sometimes tells us a great deal about the normal. We are inclined to regard people with exceptional gifts as being far removed from ourselves; however, once we start to learn a little more about how their gifts work, we suddenly find that they are more like us than we thought.

Here, two memory artists talk about the synaesthetic and eidetic perceptions that underlie their exceptional memory abilities.[1] After having seen examples in earlier chapters of the benefits that sense memories can have for people's well-being, in this chapter we look at some of the cognitive benefits. What can we learn from two memory artists to help us remember words and numbers better?

Unable to forget anything

In 1930s Russia, a neurologist received a visit from a journalist who asked him to examine his brain because he was constantly being asked by his editors how he was always able to write reports of events and interviews straight from memory without notes and without ever forgetting the smallest details and facts. The neurologist invited him to take part in a number of memory tests, in which he presented him with series of words and numbers. 'Normal' people are able to reproduce fewer and fewer words and numbers after several days of such tests; this man, however, was still able to reproduce everything days later, without omissions or mistakes.[2]

The neurologist was gripped by the unusual functioning of this man's brain, which he felt could tell him a great deal about how the human brain works. He continued testing his subject regularly; after 16 years, the journalist was still able to recall the series he had learned during the very first test. He appeared to be unable to forget anything. He eventually gave up his job as a journalist and took to the stage as a memory artist.

The neurologist, Alexander Luria, wrote a book about his years spent studying this man, entitled *Mind of Mnemonist*; the book was read all over the world and became required reading for researchers of the human memory.[3] The book offers a nice alternation between autobiographical analysis and empirical research results, and struck me above all because of its descriptions of the journalist's eidetic childhood memories and the key role that his synaesthesia plays in his memory abilities. The memory artist, whose name was Solomon Shereshevskii (dubbed simply S. in the book), visualized the things he wanted to remember. He not only made a mental photograph of the numerical series, but also linked them to personal synaesthetic perceptions. For example, all numbers were assigned set colours, while sounds evoked abstract images.

In order to remember a numerical table or series, he had to have complete silence; the tiniest sounds disrupted the process of imprinting the numbers on his memory, because the sounds appeared in his mind as 'steam clouds' and 'splashes' which made the mental rereading of the table more difficult for him.[4]

The tests performed by Luria showed that different sound pitches elicited different colours, tastes, and tactile sensations in Shereshevskii's mind. These synaesthetic perceptions of coloured and tactile sounds formed the basis for his powers of recall. Even a nonsense word or an arbitrary series of numbers acquired a colour, taste, and feel which enabled him to assign meaning to it and create a story from it. If he reproduced a word or number incorrectly, he could often tell from the colour or taste that something was wrong. He explained his synaesthesia in an interview with Luria:

> I recognize a word not only by the images it evokes but by a whole complex of feelings that image arouses. It's hard to express...it's not a matter of vision or hearing but some overall sense I get. Usually I experience a word's taste and weight, and I don't have to make an effort to remember it—the word seems to recall itself. But it's difficult to describe. What I sense is something oily slipping through my hand...or I'm aware of a slight tickling in my left hand caused by a mass of tiny, lightweight points. When that happens I simply remember, without having to make the attempt.[5]

As a child, Shereshevskii used these synaesthetic perceptions of numbers to develop a memory technique. When Luria asked him how he stored new long series of numbers in his memory, Shereshevskii answered that when he received a new instruction to remember a series of numbers, he first called up his synaesthetic arsenal of colours, tastes, and so on, and if necessary expanded it with other associations in order to create a storyline from the numerical series. He compared it to a form of stenography; rather than storing an entire word or number combination in his memory, he stored only the perceived colour or taste. This enabled him to remember far more units using fewer combinations.

Like Nabokov, Shereshevskii had lifelike and detailed eidetic childhood memories (see Chapter 7). In his childhood memories, too, fact and fiction, observation and imagination were often interwoven with each other. Luria explained this at the time by citing the ability of children to 'enchant' their environment.[6] Many children try to model the world to some degree as they would like it to be. Which of us does not remember the superstitious actions at school trying to avoid being the one whose turn it was to read out from the book or give an answer: holding tightly to the desk, thinking about a fellow pupil, and wishing the turn on them. It is a naive magic, and one which slightly twists the observable reality.[7]

Shereshevskii's memories went back far into his childhood—right back to his first year, he himself claimed.[8] Luria commented that he was unable to say for certain whether these were memories or later constructions. Right from the start, his perceptions of the environment were synaesthetically linked to other sensory perceptions, and according to Luria this was perhaps the reason that they could be preserved for so long in his memory.

> I was very young then...not even a year old perhaps...What comes to mind most clearly is the furniture in the room, not all of it, I can't remember that, but the comer of the room where my mother's bed and my cradle were. A cradle is a small bed with bars on both sides, has curved wickerwork on the under part, and it rocks...I remember that the wall-paper in the room was brown and the bed white...I can see my mother taking me in her arms, then she puts me down again...I sense movement...a feeling of warmth, then an unpleasant sensation of cold. Light is something I remember very clearly. During the day it looked like 'this,' afterwards, like 'that' —twilight. Then came the yellow light of the lamp—it looked like 'this.'[9]

Shereshevskii's recollection comes across as rather cool and lacking in emotion when compared with Nabokov's memories. But other reports of interviews with Luria show that emotion is a binding factor for Shereshevskii, too, in recalling and constructing events from the past.

> This is the sense I had of my mother: up to the time I began to recognize her, it was simply a feeling—'This is good.' No form, no face, just something bending over me, from which good would come...Pleasant...Seeing my mother was like looking at something through the lens of a camera. At first you can't make anything out, just a round cloudy spot...then a face appears, then its features become sharper. My mother picks me up. I don't see her hands. All I have is a sense that after the blur appears, something is going to happen to me. They are picking me up. Now I see their hands. I feel something both pleasant and unpleasant...It must have been that when they wiped me, they did it kind of roughly, and it didn't feel good...or when they took me out of my crib, particularly in the evening. I lie there and it feels like 'this'...Soon it will be different—like 'this.' I'm scared, I cry, and the sound of my own crying only makes me cry harder...Even then I understood that after 'this' feeling there would be noise, then stillness. Right after that I could feel a pendulum, a rocking back and forth...[10]

Finally, Shereshevskii's synaesthesia meant he did not easily forget things, but he did sometimes have difficulties in recalling them, for example if the image was not clearly visible to his mind's eye. As mentioned earlier, noise could 'splash' the image and make it more difficult to 'read'. In reality, however, he had difficulty forgetting anything. Whereas most people have to make an effort to remember things, Shereshevskii had to make an effort to forget them. He did eventually develop a technique for this. Because he visualized all his memories, he learned to forget events by screening them off behind an imaginary sheet.

Pi as a coloured landscape

Shereshevskii was never nominated for the *Guinness Book of Records*, but he would undoubtedly have been given a place in its pages. Today, there is a section in the book devoted to the somewhat bizarre ability to memorize the most digits of the constant pi. The current record-holder is the Chinese Lu Chao, who has successfully recited 67,890 digits of pi.[11] The European record-holder Daniel Tammet, a British writer living in France, was able to remember (only) 22,514 digits of pi. Tammet has Asperger's syndrome. He is also one of the few people who is able to talk about this condition, and he has described his personal experiences very clearly in his widely acclaimed first book, entitled *Born on a Blue Day* (2006). He uses a similar synaesthetic technique to Shereshevskii and also sees numbers in colours, shapes, and textures. His synaesthesia also affects his perception of words and plays a role in his work as a full-time writer who has lived in Paris for many years.

In his second book, *Embracing the Wide Sky*, he writes the following about the colours and shapes of his mental visualization of pi:

> Although the digits of Pi are, mathematically speaking, strictly random, my internal representation of them was anything but—filled with rhythmic strokes and structures of light, colour, and personality. From this random assembly of digits I was able to compose something like a visual song that meandered through every contour of my mind through which I was able to hear music of the numbers.[12]

When he thinks of pi, he sees a gently undulating mental landscape in his mind (see Figure 11.1). Like Nabokov and Shereshevskii, he has eidetic memories, which enable him to describe a picture in his memory with great accuracy and to zoom in on the landscape of pi and pick out certain colours—because it is of course impossible to see 22,514 digits at a single glance.

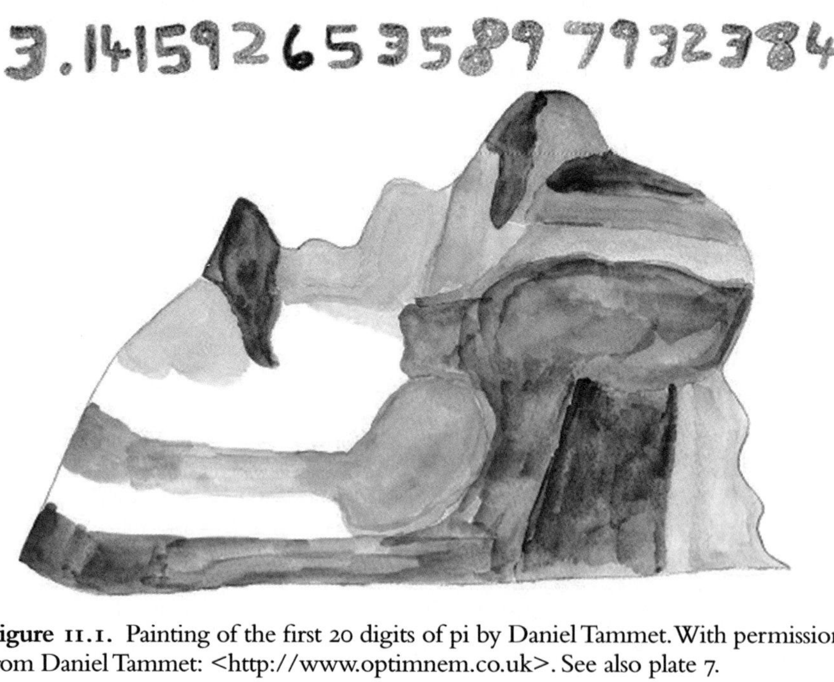

Figure 11.1. Painting of the first 20 digits of pi by Daniel Tammet. With permission from Daniel Tammet: <http://www.optimnem.co.uk>. See also plate 7.

In his third book, *Thinking in Numbers*, he recalls how his fascination with pi started in childhood:

> Long after my schooldays ended, pi's beauty stayed with me. The digits insinuated themselves into my mind. Those digits seemed to speak of endless possibility, illimitable adventure. At odd moments I would find myself murmuring them, a gentle reminder. Of course, I could not possess it—this number, its beauty, and its immensity. Perhaps, in fact, it possessed me. One day, I began to see what this number, transformed by me, and I by it, could turn into. It was then that I decided to commit a multitude of its digits to heart.[13]

He compares his method of learning the digits of pi by heart with the way a painter creates a landscape painting:

> Printed out on crisp, letter-sized sheets of paper, a thousand digits to a page, I gazed on them as a painter gazes on a favourite landscape. The painter's eye receives a near infinite number of light particles to interpret, which he sifts by intuitive meaning and personal taste. His brush begins in one part of the canvas, only to make a sudden dash to the other side. A mountain's outline slowly emerges with the tiny, patient accumulation of paint. In a similar fashion, I waited for each sequence in the digits to move me—for some attractive feature, or pleasing coincidence of 'bright' (like 1 or 5) and 'dark' (like 6 or 9) digits, for example, to catch my eye. Sometimes it would happen quickly, at other times I would have to plough thirty or forty digits deep to find some sense before working backwards. From the hundreds, then thousands, of individual digits, precisely rendered and carefully weighed, a numerical landscape gradually emerged. [14]

So that was what he finally did during the 5-hour recital of the decimal places of pi at the Museum of the History of Science in Oxford. He followed the flow and rhythm of the coloured numbers in his mental landscape painting.

Like other people with Asperger's syndrome, Tammet is exceedingly good at certain tasks, but when growing up he found some everyday tasks like playing a DVD or phoning for a taxi, more difficult. And although he knows so many digits of pi, he was actually not interested in algebra at school. On the other hand, as we shall see later, he is excellent at solving some arithmetical sums. He is also hypersensitive and is, for example, unable to obtain a driving licence, because his attention wanders at the least distraction in his environment.

Ever since he was a child, numbers have had personalities for Tammet.[15] Isolated as he was from the 'normal' human world, he was able to play with them as if they were his friends. Because he got to know the numbers as different personalities each with their own well-defined characteristics (e.g. friendly, shy) and recognizable features (colour, size), he had little difficulty in recognizing them when they were part of a numerical series—just as our friends stand out immediately among large groups of people:

> Numbers are my friends and they are always around me. Each one is unique and has its own personality. Eleven is friendly and five is loud, whereas four is both shy and quiet—it's my favourite number, perhaps because it reminds me of myself. Some are big—23,667,1179—while others are small: 6, 13, 581. Some are beautiful like 333, and some are ugly like 289. To me, every number is special.[16]

Figure 11.2 shows two drawings of numbers by Tammet.

Tammet uses the synaesthetic characteristics of numbers to remember both series of numbers (e.g. telephone numbers) and for performing arithmetical tasks. He enjoyed learning to do the latter, which felt as if he were playing with his friends. He developed his own method of visual thinking and remembering. Although he was not particularly good at algebra at school, he is able to perform certain complex arithmetical tasks with striking ease.

> My favourite kind of calculation is power multiplication, which means multiplying a number by itself a specified number of times. Multiplying a number by itself is called squaring; for example, the square of 72 is 72 × 72 = 5,184. Squares are always symmetrical shapes in my mind, which makes them especially beautiful to me. Multiplying the same number three times over is called cubing or 'raising' to the third power. The cube or third power of 51 is equivalent to 51 × 51 × 51 = 132,651. I see each result of a power multiplication as a distinctive visual pattern in my head. As the sums and their results grow, so the mental shapes and colours I experience become increasingly more complex.

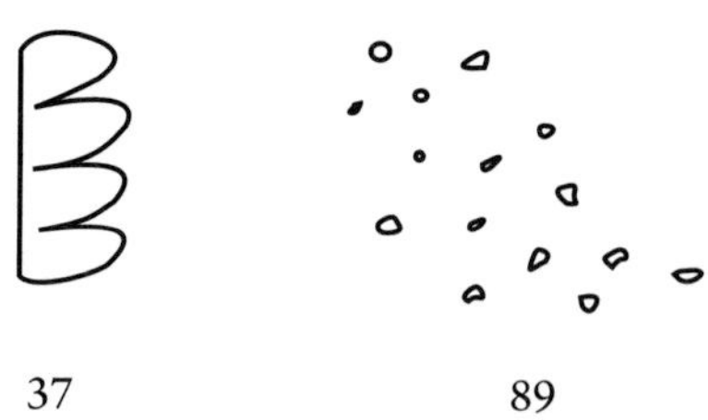

Figure 11.2. Drawings of the numbers 37 and 89 by Daniel Tammet: '37 is lumpy like porridge, while 89 reminds me of falling snow'. With permission from Daniel Tammet: <http://www.optimnem.co.uk>.

> I see thirty-seven's fifth power—37 × 37 × 37 × 37 × 37 = 69,343,957—as a large circle composed of smaller circles running clockwise from the top round. When I divide one number by another, in my head I see a spiral rotating downwards in larger and larger loops, which seem to warp and curve. Different divisions produce different sizes of spirals with varying curves. From my mental imagery I'm able to calculate a sum like 13:97 (0.1340206...) to almost a hundred decimal places.
>
> I never write anything down when I'm calculating, because I've always been able to do the sums in my head and it's much easier for me to visualise the answer using my synaesthetic shapes than to try to follow the 'carry the one' techniques taught in the textbooks we are given at school.[17]

Tammet's first memories of coloured numbers go back to when he was aged 4:

> The pages of my books all had numbers on them and I felt happy encircled by them, as though wrapped in a numerical comfort blanket. Long before I could read the sentences on the pages, I could count the numbers. And when I counted, the numbers would appear as motions or coloured shapes in my mind.[18]

Whilst at school, he discovered that the school system was not equipped to deal with his notion of numbers and that he performed calculations differently from his classmates:

> I often found it confusing when we were given arithmetic worksheets in class with the different numbers printed identically in black. To me, it seemed that the sheets were covered in errors. I couldn't figure out, for example, why eight was not larger than six, or why nine wasn't printed in blue instead of black. I theorised that the school had printed too many nines in their previous worksheets and had run out of the right colour ink. When I wrote my answers on the paper the teacher complained that my writing was too uneven and messy. I was told to write every number the same as the others. I didn't like having to write the numbers down wrong. None of the other children seemed to mind. It was only in my teens that I realised that my experience of numbers was very different to that of the other children.[19]

Tammet's exceptional memory abilities attracted the attention of scientists, and he received an invitation from a group of synaesthesia researchers at the neurological laboratory of the University of California in San Diego. To test Tammet's synaesthesia, the researchers presented him with sheets containing 100 numbers and asked him to choose colours, textures, and shapes for each number. When repeating the numbers, Tammet linked the numbers to these specific colours, textures, and shapes very consistently. While members of a control group with normal memories were able to remember an average of only around four of the 100 numbers after one day, Tammet could still remember 68 of them. The series of numbers had evidently become engraved on his memory, because after 3 days he was still able to reproduce all 68 numbers.

Like other synaesthetes who see numbers in colours, Tammet's performance deteriorated when he was asked to remember series of numbers whose colours differed from 'his' number colours.[20] After 3 days he was not able to reproduce a single one of them, compared with the 68 he had been able to reproduce after 3 days in the previous test.[21] Research on Tammet's brain later showed that different parts of his brain were active when remembering numerical series than in the brains of members of the control group without synaesthesia, suggesting that his brain was following a different strategy.[22]

Do synaesthetes have better memories?

Synaesthetes often say that one of the advantages of synaesthesia is that they can easily remember names and telephone numbers by using the colours of the letters and figures. Both Shereshevskii and Tammet used their synaesthesia in this way as a technique for remembering. Moreover, their synaesthesia elicited eidetic memories, for example, producing detailed images which they were able to search mentally. Do other synaesthetes also have this ability? Do all synaesthetes have exceptional memories?

Dan Smilek and his colleagues at the University of Waterloo investigated the exceptional memory of a synaesthete for whom numbers had set colours.[23] She had little difficulty in learning lists of numbers off by heart, according to her own testimony using the colour combinations. Even after 2 months she was able to reproduce the lists without difficulty.

The researchers then subjected her to a test comparable to that carried out on Daniel Tammet in San Diego. They asked six non-synaesthetic volunteers to take part in the same experiment to see whether she really did perform better than other people. The subjects were given three different series of numbers in three sessions. In the first session, the numbers were printed in normal black ink on white paper; in the second, they were printed in colours on white paper; and in the third session the numbers were printed in colours which corresponded with the colour-number combinations of the synaesthete.

In the first session, with the black figures on white paper, the synaesthete performed much better than the other participants. In the session with the arbitrarily coloured numbers (which were deliberately printed in colours that did not correspond with her number–colour combinations), she performed worse than the others, because she said the colours got in the way and distracted her. In the third session, where the numbers were printed in 'her' colours, she was once again the best performer. Strikingly, the other participants also performed better in this session. Perhaps they were also benefiting from the 'synaesthetically' coloured numbers. In any event, they performed better than in the session with the numbers printed in arbitrary colours.

A similar result was found in an experiment involving another synaesthete who saw letters in colour. She was tested by Carol Bergfeld Mills and her colleagues from Goucher College in Baltimore, in the United States. They found that she was much better at remembering lists of words than non-synaesthetes. In other, non-linguistic memory tests, however, such as one involving illustrations, she did no better than the other participants.[24]

Can these results be put down to coincidence? Did these researchers just happen to find synaesthetes with good memories, purely by chance? To rule out this possibility, Caroline Yaro and Jamie Ward from University College London studied a large group of synaesthetes who perceived letters or numbers in colour.[25] They compared 46 synaesthetes with 46 non-synaesthetes using a battery of varied memory tests, each of which tested different aspects of the memory.

Generally speaking, the synaesthetes did slightly better on the battery of tests than the non-synaesthetes. The researchers also found a few synaesthetes with exceptional memories, which were reminiscent of the amazing memory of Shereshevskii. Five out of the 46 synaesthetes had what the researchers called a 'superior memory'.

On the other hand, there were also synaesthetes with a poorer than average memory. The synaesthetes performed better mainly in the tests for colour memory and visual memory, but did worse in remembering groups of numbers, for example. According to the researchers, their advantage lay in the double encoding of the numbers: in addition to the number itself, the associated colour was also imprinted on their minds and they therefore had twice as many cues to help them reproduce the matrices.[26]

The psychologists Nicolas Rothen and Beat Meier from the University of Bern, Switzerland also compared the memory performance of synaesthetes who associated letters with colours with a group of non-synaesthetes who were comparable in terms of age, sex, and education level. They found no differences between the two groups in a test involving visual search tasks and a memory test.[27] The researchers suggested that it is not so much the synaesthesia as the strategic use of synaesthesia that makes the difference in the ability to remember names and events. Evidently, some synaesthetes have found a strategy for remembering things better.

These results add some perspective to the notion that synaesthetes have better memories than non-synaesthetes.[28] Some synaesthetes do indeed exhibit a better memory performance, but there are also synaesthetes who score worse than non-synaesthetes. Synaesthetes do better mainly on tasks in which they can use their letter–colour or number–colour associations. This suggests that synaesthetes may not so much have better memories than other people, but that some of them are able to use their synaesthesia to develop memory techniques and strategies which enable them to remember certain things (words, numbers) better than others.[29] Shereshevskii and Tammet have then exploited this ability to the full.

Orientation in memory

Experiencing synaesthetic or eidetic perceptions is not the same as having a good memory. A great deal depends on how individuals develop these sensory abilities into memory techniques. People mostly develop these techniques during childhood, as the autobiography of Daniel Tammet illustrates.

The fact that one person becomes a memory artist and another does not has to do with more than sensory abilities alone. There is great variation in the intensity and nature of synaesthetic and eidetic experiences. Shereshevskii and Tammet have very intense and multiple forms of synaesthetic and eidetic abilities, as a result of which they had a wide range of memory cues at their disposal as children. As a result, during their childhood years they subconsciously built up a bigger (syn)aesthetic capital of sense memories than other children (see Chapter 8).

A great deal obviously also depends on the individual life-course of the person concerned. Would Tammet have become a memory artist and a writer if he had not had Asperger's syndrome and had chosen human friends over his numerical friends as a child? The German anthropologist and synaesthesia expert Alexandra Dittmar highlighted the individual differences in the 'orientation' of synaesthetes.[30] Orientation is a term that Dittmar has borrowed from anthropology, where it refers not only to mental representations which enable people to find their way, but also to collective representations or world images which give direction to people's actions. An example is the right–left orientation which is familiar to most people. Left and right help us to orientate ourselves from the moment we get out of bed (with our left or right leg), turn the handle of the bathroom door (with our right or left hand), and so on. This left–right orientation is so natural that we are barely aware of it. However, there are also cultures where people, instead of a left–right orientation, apply a 'next to each other' orientation.[31] Children who have not yet learned to differentiate between left and right also use other orientations to make sense of their surroundings and find their way around them. Although they manage to do this without having a left–right orientation, once they do have that orientation it becomes easier.

Synaesthetic abilities can be regarded as a useful set of tools which help synaesthetes organize and find their way around their surroundings. They are stable, consistent, and therefore provide an excellent footing. Synaesthetes report that their synaesthetic perception can help them to understand and get to grips with new situations, for example, because the colour associations enable them to organize the bombardment of new impressions at lightning speed, often subconsciously and before the synaesthete themselves is aware of what they are doing.[32]

Orientation is part of the memory.[33] When a person who perceives numbers in colour is presented with a new calculation task, the figures in the sum automatically arrange themselves as coloured elements in a 'hall of numbers'. Then the calculation can begin. Tammet, for example, uses coloured numerical landscapes to reproduce thousands of digits of pi. Other synaesthetes often first remember the colours of a piece of music that they have heard, and only later remember the sounds.[34]

Orientation naturally has to do with space. Some synaesthetes are able to navigate through a mental hall of numbers, a room of coloured years, months, and days of the week or a coloured memory of a concert they have attended. These synaesthetes are also aware of their own position in that space. They occupy it with their body and are able almost to feel their surroundings with their senses. The advantage of a synaesthetic orientation is that it is not readily forgotten, because it is part of the synaesthete's everyday physical experience. This explains why these synaesthetes do not easily forget certain impressions and have better memories than other people.

This applies not only for remembering series of numbers, but also for autobiographical facts. Researchers Julia Simner and her colleagues at the University of Edinburgh carried out a study of a group of ten synaesthetes who experience time and their life-course as if it were something that surrounded their bodies, for example, seeing a circle of years made up of coloured months, with the red January in front of them and the orange July behind them. They compared the performance of this group of synaesthetes with a randomly selected control group of people without synaesthesia. The synaesthetes remembered approximately twice as many autobiographical facts as the non-synaesthetes. And even though they were not factually any more correct, it was clear that their estimations of the days on which historical events had taken place were more accurate than those of the control group.[35] Synaesthesia thus helps people orientate themselves in time and space.

Children would benefit if their synaesthetic and eidetic orientation in learning to add additional characteristics such as colours and personalities to numbers and objects (see examples in Chapter 8) were to be encouraged at school.[36] That would anchor their lessons and other important life impressions more firmly in their memories. They would also continue to derive pleasure in later life from that broader access, as the accounts in Chapter 12 will show, as we learn how synaesthetes colour their past.

12

How people colour their past

Synaesthesia or how the senses colour present and past

Synaesthetes such as Daniel Tammet and Solomon Shereshevskii in Chapter 11 add colour to numbers and words in the present in order to remember them better. Do synaesthetes also colour their past? To what purpose? What benefit might synaesthetes derive from this? That is the subject of the accounts presented in this chapter. We will see that what we sometimes describe metaphorically as 'colourful memories' or 'a colourful past' can be applied very literally for synaesthetes.

I placed an advert in the Dutch synaesthesia newsletter *Synesthesie Nederland* asking for people to take part in a study of synaesthesia and childhood memories. Lots of synaesthetes responded. They completed a questionnaire on this topic and, based on the responses, I interviewed a number of them in more depth about the role that synaesthesia played in their childhood memories.[1]

Coloured sounds of a flood disaster

During my conversation with coloured-letter synaesthete Janneke van Dijk about her childhood memories, something unusual happened. Her sense memories transported her back in time and she re-experienced aspects of the past that were partially new. For example, she was struck by a memory of the night of the 1953 flood disaster in the southwestern Dutch province of Zeeland; the memory began dark and shadowy, before colour was suddenly injected by the people and the coloured names of those who were around her at the time.

Janneke was 8 years old at the time of the flood disaster. From a very early age, she was aware that she heard names in colour:

> From about the age of three. One clear memory of that time is of telling my mother that Auntie Sjaan, which was what we called our neighbour, was light green in colour with gold flecks. But at that time I saw the sound of her name as a field of colour, a living area, not as coloured letters.[2]

When she learned to write, the sound colours of names changed into letter colours. Names, and therefore people, acquired the colour that emerged via the letters.

> When my mother said a word, I saw colours when hearing its sound; if she introduced someone by name, I saw that person's name in colour. [...] From the time that I was able to read and write, the colours of the letters took the place of people's names. When I think back to that time, I see the colours of those names shooting very fast through the memories. It's as if I'm reading the colours of the names. [...] I see the names written out in front of me.[3]

Her synaesthetic perception of coloured names helps her to construct sense memories, as for example when she hears a storm outside her home in the Dutch town of Wageningen, evoking memories of the night before the 1953 flood disaster:

> Whenever there's a storm like on the night before the flood disaster, it evokes associative feelings for me. Although I really like storms, the crashing and banging can also make me restless and nervous [...]; I'm then gripped by a feeling of powerlessness, a sense that something is coming that I can't do anything about, but that I can also watch as a sort of bystander.[4]

The colours of her parents' names appear in her memory of that night:

> My father's name is Piet van Dijk. He's a sort of light blue colour, and that colour flashes by when I remember that he wasn't there. Yes, that comes very quickly. My mother has a colour that's hard to describe, something along the lines of a mix of brown, pink and a sort of orange. She was there, and that colour therefore flashes past in my memory when I think of that night. My little sisters and my brother, the neighbours who came to stay in our house, they all have their own light, living energy in the middle of that event.[5]

When I asked her whether the colours of her present and absent mother and father are also associated with emotions in her memory, she replied:

> Yes, the absence of my father and his colour in my experience of that night were associated with a feeling of anxiety, loneliness at that moment when he wasn't at home; he was on the dyke to lay sandbags.[6]

This memory is evoked by the sound of the howling storm around her house and, through the feeling of powerlessness, she is led to images of that night. The colours of the names of her present mother and absent father shape the images in her memory (Figure 12.1).

Other sensory stimuli evoke pleasant childhood memories in the Zeeland countryside: the warmth of the sun, the smell of wheat in the fields, the crackling of cat ice, the crunch of breaking seashells, the smell and warmth of cows' bodies and soft udders. The smell of the wheat alone is enough to trigger a feeling of joy and intense, living, tangible sense memories:

> I can smell that smell of happiness again now, of being carefree, being free. [...] Wheat brings back the sound of the wind, and the smell and sound of a threshing machine. The sound of the wind and the glittering sun on the stalks. The colours are very vivid; they thrill and vibrate. It gives me a feeling of intense happiness.[7]

Figure 12.1. Janneke van Dijk, February 1953. *When the waters came*... She writes about the work:
'It is a chaotic image, as if the ground is falling away beneath me. The blue above is the colour of my father and the brownish tint below is the colour of my mother'. © Janneke van Dijk. See also plate 8.

When I asked her whether those childhood memories would have been different without her synaesthesia, she answers in the affirmative: 'Less full, less complete, less "everything at once". Yes, I have the feeling that synaesthesia gives them a unity that would otherwise not be there'.

The memories of the sound of the floods and the images and smells of wheat waving in the summer breeze merge into a single image. The emotions of both powerlessness and happiness are each given a place in the collages that her memory constructs from her senses.

Memories of the skin

Synaesthesia also colours the memories of Elsbeth Rabus. When she thinks back to her time at primary school, she sees images, and as soon as people enter those images, their names appear in colour. She knows that she began seeing colours associated with names at around the age of 4, shortly after moving to the Dutch town of Apeldoorn:

> I remember playing with other children in the Loolaan in Apeldoorn; I can still see precisely where that was. I'd invented a game. I got the other children to call out their names and I told them what colour their name was. They liked that, but when I asked them what colour my name was, the game stopped and I discovered that other children didn't see what I was seeing…that I was unusual…[8]

Her earliest memory goes back to the age of 2, and she can still see it clearly:

> I see a royal blue glazed milk beaker, shiny, with a white inside: it's the loveliest blue I've ever seen and I've never been able to find it again since. I also know that the memory doesn't come from a photograph, because no one ever took a picture of me with that blue mug, not even in black and white. That blue mug is still as beautiful and shining blue as ever. I would recognise the colour again instantly, but I haven't managed to find it yet [laughs]. I know that I was two years old: Cologne was suffering heavy bombing and my grandmother took me to the south of Germany. That's how I know how old I was at the time.[9]

The smell of beer also takes her straight back to her life as a 2-year-old. She still has very clear sense memories. It is not just the smells, sounds, and images, but above all the positions of her body and the tactile impressions that awaken the memory:

> Whenever I go into a bar that smells of beer, it triggers a memory of a German restaurant. I was around two and a half years old. I can see myself climbing

> up onto a high stool and the way the drips falling out of the tap on to the shining metal bar top. I can still remember precisely the movements my body made and how I had to move my head to catch the drips from the tap on my tongue.[10]

Elsbeth followed a course in textile arts in Denmark and has what she calls a strong 'memory of the skin'. She produces a list of examples: 'cold winter toes; scabs on your grazed knees; swollen, itching gums that you can't keep your tongue away from; bare feet on stone paths or prickly, dry grass; the feel of your cheek against a moleskin coat; scratchy woollen stockings reaching halfway up your winter-chafed thighs; a dress made of parachute silk (rustling and hard; see Figure 12.2); and the merciless woollen blanket when the sheet has slid to one side':

> If I see a child fall over, grazing its knees, I feel the pain of that damaged skin and then I see myself with bleeding, dirty knees. It was wartime, I ran, fought and fell over a lot, and it always got inflamed and developed interestingly coloured scabs; after a while I could carefully pick and scratch at them without it hurting too much, and I was always amazed to see the glowing new pink skin underneath.[11]

In Elsbeth's memories, images and tactile impressions almost always go together:

> On my sixth birthday I was given a doll in a light blue four-poster bed, with a canopy made of very soft material in a clear sea-green colour. Whenever I come across that material I can recall that doll and the four-poster bed. It's not just the colour, but also the smoothness of the artificial silk, the touch of the material; that's my 'textile memory', I guess. That was the first time in my life that I'd felt a material like that.[12]

Recolouring emotions

The visual artist Clara Froger, with whom I organized several painting workshops for synaesthetes and children, was the first person to make me aware of the synaesthetic colouring of the present and past.[13] She sees colours in association with sounds, smells, and tastes and other sensory impressions and emotions. She has had this ability all her life. During lectures and workshops, she uses this synaesthetically coloured environment as an extra sense that enables her to listen, perceive, and remember more carefully. The colours help her to organize information.

Figure 12.2. Childhood photo from 1949 of Elsbeth Rabus as a 10-year-old in a Sunday dress made of parachute silk, which in her tactile memory felt unpleasantly rustling and hard. With permission from Elsbeth Rabus.

In her work as a visual designer she not only uses colours as visual information, but also the acoustic, tactile, and olfactory qualities of the colours as they appear in her perception (Figure 12.3):

> They are just like paintings, but then abstract coloured images that are in motion, very spatial. The colours of sounds seem to float in the space. But that space also seems to be inside my head. When it happens at concerts, I remember those concerts really clearly. When I think back to them, I see only the coloured images; I don't hear the sounds. I see the coloured images of the orchestra, but don't hear them. I 'know' the sounds; I can name them, but I don't 'hear' them. I only experience the coloured images. By contrast, if I make music myself, such as singing the St Matthew Passion, it's the sound

Figure 12.3. Clara Froger, *Not sleeping on the northern Rotterdam ring road*, 2009. She explains: 'Colour images that pass by in constant motion. Lorries and cars, trains in an unending stream of sound and colour and the alarm clock as a dotted line'. © Clara Froger, 2009. See also plate 9.

that sticks very firmly in my mind. Whilst we're singing, though, I see the colours as well, so I get double the pleasure![14]

Clara has been colouring her sensory impressions from a very young age. Like Elsbeth Rabus, Clara has a detailed tactile memory:

> From when I was a small child, I've always wanted to touch everything. It's an urge I still have today [laughs], so I do it often. [...] It's quite possible for me to touch a white surface, say, but that a different colour comes into my head. The whiteness of the surface has no influence on the colour that's in my head; that's determined by the temperature of the surface I'm touching. A hard, cold surface produces a different colour from a warm surface. [She touches the terracotta-coloured marble top of the café table where we are sitting.] This material brings back memories of our house. We had quite a lot of marble in the house.[15]

When asked about the colour of tactile memories, she explained that when she touched the cool, dark-brown marble table-top, its temperature brought back memories of white marble walls in her parental home, because 'cool is often white'. The emotions are never far away. Why does the tactile impression take her back to an emotionally charged environment? What role do emotions play in these coloured tactile memories? Her answer was as follows:

> Difficult and emotional memories are situated in a particular colour zone. If I notice that I'm being drawn into that zone, I also feel that I don't want

> to be there. It helps if I eliminate that colour. It might sound odd, but I had quite a complex childhood. And then it's important that I can eliminate the colour of a particular moment. It makes it easier for me to leave that memory. Strange, how I do that. I then try to escape from the memory by eliminating that colour.[16]

She became visibly emotional, and I asked her if the same thing happened with pleasant memories:

> Yes, but then I leave the colour alone! [laughs]. But yes, it works in just the same way, really. Actually, that would be a nice form of psychotherapy: helping people to colour their memories. Painting memories, events, seems to help lots of people. I was already doing it as a child.[17]

She recalled that as a child she was already using colours to deal with emotions:

> Dark, velvety red is always negative for me. It signifies memories of negative events. Happy memories are always transparent, very light and purer in colour. In difficult situations I also called up colours to help me. I was already doing that as a little girl of five. When things got difficult, I 'cried' myself into a colour. [...] That was coming from inside, of course. [...] If the situation was dark red, I tried to put myself in a different colour.[18]

Colour as a mental cocoon

Clara Froger is not the only person who recolours her memories. In my interview with Nelleke van den Oever, similar forms of coping emerged. Nelleke had a career as an actress and now gives drama lessons in secondary schools and to autistic children. She said the following about her synaesthesia: 'I see letters, words, days of the week and above all numbers in colour. I also see colours around people and objects, and I can also surround myself in colours, for example to make myself feel better'.[19] As an actress she never had any difficulty remembering scripts, because they appeared as coloured subtitles in her mind's eye. Her synaesthesia means she has a good memory for conversations because she can see the same coloured subtitling of those exchanges. She also spent some time working for a newspaper and remembers that she only needed to write three words when reporting an interview; the rest of the conversation then simply unfolded as if by itself:

> I see words before me and they contain colours. When I recall a conversation, I see the text above it, appearing like surtitles. The letters don't have a specific

> colour. [...] People sometimes say: 'Gosh, it's amazing that you still remember that; that's exactly what I said!' [...] The rhythm gives me something to hold on to; it's as if I see the commas in the sentence. [...] I see colours associated with it; I see the people in front of me as if in a picture.[20]

She remembers coloured events better than events and objects without 'colour', which she forgets fairly readily:

> The bigger the impression an event has made on me, the more completely it's coloured in. For example, I can still recall precisely how the light came in through the window of my room as a child. [...] Colour can take me back into a memory. Purple always made me think about my childhood girlfriend in the 1970s. Whenever I come across that colour, it takes me straight back. Then I can remember images, light and dark and sounds.[21]

People can also have an aura of colour in her memories. The colour depends on the energy of the person concerned. As an example, she describes theatre performances by her favourite actor as being very alive and colourful:

> Sometimes, a person will be surrounded by blue. Certain charismatic actors have an amazing aura of light, very strong, very white, very bright; I see a complex colour structure made up of energies. It knocks me sideways. Other people may not see those colours, but they still feel the power, I think. I then feel a desire to be as close to that person as possible.[22]

Since her childhood she has been sensitive to people's colour energies. Colour played a role for her in the same way as intuition and emotion. As a child, she was able to place herself 'inside' a colour.

> If I don't feel well, I feel the need to colour in pictures. I enjoy doing it and it makes me feel better. [...] I can also surround myself with colours. I suffered from stress for a while and I was so tired that I had nowhere left to go, and then I surrounded my body with a sort of mental cocoon of red and white, a sort of helical shape that I could crawl right into, and I felt good there—in my head, in my thoughts. [...] If I don't feel well, I have to wear red, because that protects me.[23]

Colouring seasons

When I interviewed the synaesthete Martina Strusny, she was a student at the Gerrit Rietveld Academie in Amsterdam.[24] As a child, she grew up in the former German Democratic Republic, and moved to the Netherlands 6 years ago. She sees letters, words, and numbers in colour when she thinks about them and when people speak to her slowly. She sees months, weeks, and days as three-dimensional coloured shapes. Each month also has its own specific feeling. She became aware of her synaesthesia after a fellow student told her that there was a scientific term for the unusual way in which she perceived and remembered things. After this, she researched her 'synaesthesia' with a passion and made it the subject of an artistic experiment for her graduation project at the Rietveld Academie.

It turned out to be an emotional adventure. She told me that her sense memories caused her to swing back and forth between happiness and nostalgia. Growing up in East Germany, she lived through the revolution and subsequently moved to the Netherlands. Moving home and all the changes in her life were expressed in changing smells, images, tastes, and sounds.

Her autobiographical past is built around the months of the year, which have fixed sensory characteristics for her, especially colour. January, for example, is white, and this colour also contains associations with the cold winters in East Germany. She perceives the months as collages of associations that are linked with emotions from her youth. After moving to the Netherlands, the colours of the months no longer made sense, she explained. Dutch winters have little snow and also lack the emotions that she felt in the East German winters. The colour differences between the months are less sharp in the moderate maritime climate of the Netherlands than in the more extreme continental climate of East Germany:

> If I look around in my memory from June or July onwards, the winter months are darker and situated lower in the image. I see candlelights from my childhood, but that's changed. When I came to the Netherlands, something changed and I didn't know what it was—until I discovered that the image of the winter had changed. January and February were no longer white, but had become grey, and filled with small yellow dots of narcissi. The colours of the months changed gradually because of changes in my life.[25]

Each month evokes a different collage of images and emotions for her:

> April is yellow and green and also a little bit grey, because there are still frequent winter storms with cold rain showers. The letter 'a' is yellow; there is sun and there are also yellow flowers in this month. The months become clearer from March onwards, and from the autumn they start becoming darker again. I see the written names of the months in colour. But it's more a collage of memory fragments mixed with the coloured names of the months. They are mainly images, not sound or tastes or smells [...] June is pink and light blue; July is the colour of brambles and tall grass, and August is yellowy

> orange-brown, like the fields at that time of year. In July and August I can even hear the heat (hot summer air and silence). The memories in images are very intense; I'm present in them and I even feel heartbeats.[26]

She can step into a memory, as it were, helped by her synaesthesia:

> My memories are situated on a long timeline. I can follow that timeline in my mind and go to a particular branch, which then becomes more prominent, and then I can see everything from that time. I experience it like the day before yesterday...intense, and some details are clear. I can literally sit there, be there and look around.[27]

After a period in which the synaesthetically coloured memories caused confusion and mixed emotions, she was able to transform her synaesthetic perception of past and present into positive feelings. She later explained that she had found an image form which had taken her to new kinds of happiness:

> For the last few weeks I've been working on an art project based around the time when I first discovered my synaesthesia. The fact that there was actually a term for seeing letters and numbers in colour, and that from that point on I was able to say that to everyone loudly and clearly, make me so unbelievably happy and cheerful that I wanted to hold on to my first impression.
>
> In the same way that a bookkeeper likes their numbers to be in proper order, or like a child playing with a brightly coloured gift they have received spends a long time looking at it from all sides, shows it to everyone, even takes it to bed, that's how I played with my synaesthesia.
>
> Later, I designed a mental scale, as well as a physical one, which weighs all warm letters and numbers (red, orange, beige, yellow,...) and all cold ones (green, blue, grey,...) against each other. I made all letters that have no colour white, and they simply lay on the ground. Now I could see how much of everything I had. And I saw that the warm side was weighed down much more heavily.
>
> The letters and numbers are made of wood and very brightly painted with oil paint and acrylic (see Figure 12.4). My mental scale is made from moulded glass, because the clarity and secretive consistency of moulded glass comes very close to the world of thoughts.[28]

Synaesthetic colouring of the past

The stories told by these synaesthetes show that they colour their past in a very special way. Most people are able to recall personal events from their

Figure 12.4. Martina Strusny, *Viel mehr Rot als Blau* ('Much more red than blue'). Photo: Ron Zijlstra. © Martina Strusny, 2013. See also plate 10.

past and to describe them in terms of colours and other sensory qualities. Synaesthetes, however, literally see those colours before them, in perceptions and memories. They are so palpably present that they are able to describe or paint them in detail.[29]

The accounts in this chapter show how synaesthesia lends colour and shape to the past of synaesthetes. Their synaesthesia seems to crystallize the emotions in childhood memories. Janneke van Dijk's memory of the night of the 1953 flood disaster is emotionally coloured by the brown, pink, and orange colour of her mother and the light-blue absence of her father. These synaesthetic colours give the emotions tangible form.

Their synaesthetic perceptions appear to organize the sensory fragments of the past and unifies them. The smell of wheat transports Janneke back to the fields of her youth and evokes a variety of sensory impressions. She believes that without her synaesthesia her memory would be incomplete.

Synaesthetic perceptions bring the past to life. The joining in of the senses makes it easier for synaesthetes to identify with the events. This empathic experience of the past is sometimes so strong that the synaesthete has the feeling of being physically present in it.[30] Elsbeth Rabus had the physical sensation of being back in her childhood body at the age of 2 and could feel how her body moved or how she bent her head to catch a drip from the tap.

The synaesthetic colours offer a tool to cope with dramatic events from the past. Both Clara Froger and Nelleke van den Oever sought emotional protection by withdrawing into colour. These coloured memories offer an additional perspective on autobiographical narrative-building that is often studied from a lingual perspective.[31] In addition to verbal narratives, sense memories can be explored as a way of understanding and coping with emotional problems of the present and the past.

Moreover, interviewing people with synaesthesia offers an insight into the creative process of composing the past as a collage. For example, Clara Froger recounted how the colours of the marble walls in her parental home are white in her memories, an emotionally cold white, whereas in reality they were probably different colours. Martina Strusny described how her brain appears to rearrange the collages of her memories and colour them differently. These are examples of how sense memories help in constructing an autobiographical past and imbuing it with different colours.

Whereas for most of us the 'coloured past' is something we can only describe figuratively, for synaesthetes it is literal and physically palpable. How can non-synaesthetes become more aware of the sensory colouring of the past? Can sense memories be put to use at all, or do they simply happen, as Proust wrote, and can we only wait until they suddenly appear? That is the topic of the final chapter.

13

Enjoying sense memories

Concluding remarks

Most of our memory activities in daily life—and most scientific studies on memory—are focused on the controlled retrieval of facts, while many of our everyday memory experiences are non-deliberate and often elicited by accidental sensory stimuli such as a smell, a taste, a sound, an image, or touching something. Following on from Proust, sense memories have been investigated and explored principally in the arts disciplines, though there is also a modest but growing body of attention for sense memories in a number of disciplines such as health care and education. The number of projects focusing on the senses and memories is increasing particularly rapidly in the field of elderly care.[1] Bringing together these insights from science, the arts, and practice, I will sketch the outlines of an idea here of how sense memories might work and conclude with a reflection on what sense memories can mean for us in our daily lives.

Sense memories are characterized by the interplay of different senses. Although we often think that one smell evokes a childhood memory, what is actually happening is that several senses are cooperating to construct a collage from different autobiographical episodes. In an attempt to get a grip on this, I would venture to describe them as follows. Sense memories are memories that are triggered involuntarily by a single sensory stimulus and which in collaboration with other sensory processes in the brain create an emotionally charged and physically palpable collage which gives a meaningful but coloured representation of an episode from the personal past. It is no coincidence that Marcel Proust compared sense memories with the creative process of writing or composing. It is as if the brain makes a painting or collage which looks convincing, but in which it is no longer possible to determine precisely what is factually accurate and what is a product of the imagination.

Different places, different emotions

Sense memories open doors to our personal past that are normally locked. Which of your friends were at your sixth birthday party? You probably don't know any more. There are lots of facts that we appear to forget and lose forever in our memories. However, by using a circuitous route we can still discover who was at that birthday party, by allowing our senses to offer us a helping hand. For example, the taste of a birthday cake, the smell of the candle smoke, sound recordings of a hoarsely sung 'Happy Birthday to You', and of course a photograph of the birthday boy or girl, can bring back fragments of memories from that time or even from that day. With a little luck, as those fragments gradually connect to form more coherent wholes, you may find that voices and faces of your friends start to emerge in your memory.

This example shows that involuntary sense memories operate differently from the more common voluntary, goal-directed memories that are driven by verbal cues (e.g. asking questions, telling stories).[2] When you are trying to recall those friends at your birthday party, your voluntary memory will think about who your friends were at that time from the neighbourhood and your school and try to deduce which of them could have been at your birthday party. But it remains guesswork. It is only when 'living' memories with faces, voices, songs, and other sensory impressions appear that you feel able to say with much more certainty who actually was at your party. And of course, since a memory is a collage, it is always possible that your memory is simply 'adding' a friend who was not present at your party but at another party.

A nice characterization of the difference between verbal and sense memories was once given by the memory psychologist Daniel Schacter, who said that 'knowing' that you have been somewhere with someone is not the same as 'remembering' that you were there together.[3] Verbal memories are generally abstract and cognitive/logical in nature, focused on finding the truth. Sense memories are more tangible; they are filled with sensory impressions and emotions and create the feeling of being able to physically re-experience an event from the past. It is as if we are briefly standing with one foot in the past and the other in the present.[4]

The striking thing is that the memory of the senses does not submit to voluntary verbal commands (e.g. questions). The faces of the friends at the birthday party, to return to our earlier example, only appeared after some time through the chance hearing of a children's song. The birthday question was a conscious instruction to the memory. However, sense memories often do not work to instruction; even if you were to taste the birthday cake, this does not guarantee that the memory will come back, while at another unexpected moment, the absent-minded mopping up of cake crumbs with your finger and putting them in your mouth without really thinking about it can bring back a vivid memory, with all its colours and smells.

It has been suggested that some childhood memories are very hard to access using verbal goal-directed retrieval. In particular, according to clinical therapists early highly emotional and sometimes traumatic memories are sometimes only opened by situational cues.[5] As stated earlier, only a small number of scientific studies have been devoted to involuntary sense memories, and due to the scientific methods of written diaries and verbal retrieval, most of the knowledge gathered relates to verbal memories, leaving the area of sense memories relatively unexplored. More scientific studies on involuntary sense memories are needed to validate the findings of artists and practitioners that sense memories are very useful keys for opening lost memories.

One thing that often helps is taking the time. Sense memories do not pop up like answers to a quiz question in a couple of milliseconds. Think of the Flemish woman in Chapter 9, for example, who after smelling Peru balsam became aware only more than a quarter of an hour later of a memory of her mother giving her a cough medicine. The childhood memory of the Groningen man in Chapter 6, which was triggered by smelling sugar beet, also shows that sense memories need time and cannot be forced to produce rapid, ready-made answers. Sense memories are rarely 'flashbulb memories' which produce a complete memory image in the brain in a lightning flash; more commonly, they are gradual revelations of the past, building up slowly from fragments of sensory impressions which gradually come together to form a collage.

Emotions establish strong connections in the mental to-ing and fro-ing between past and present in sense memories. In particular, smells can evoke strong emotional links with the past.[6] But musical sounds, a touch or an old, yellowed photograph can also move us, arouse our emotions—a word which is itself related to the motion of our moving bodies. Sense memories are almost always physically palpable. It is almost impossible to build a smell memory, for example, in a remote and rational way, without becoming physically and emotionally moved by the experience.

This sensual way of remembering is sometimes called nostalgic. However, I think the memory of the senses is about much more than nostalgia. Nostalgia has something of a negative connotation. The word was coined as a medical condition by the Swiss physician Hofer in the seventeenth century. Nostalgia is made up from the Greek words *nostos* (return) and *algos* (pain) and refers to a painful state of hankering after things past.[7] On the one hand, sense memories can indeed evoke nostalgia because a person wants to stay in the past. On the other hand, the vague hankering for the past is made tangible by sense memories. Nostalgia and sense memories do occur together, but they are not the same thing. Nostalgia is a conscious desire, while sense memories are involuntary perceptions. Nostalgic feelings place us almost entirely in the past, whereas in sense memories both past and present are involved and briefly coincide, as Marcel Proust suggested in his reflections on the subject.

Different brain pathways

Although a single sensory impression can set in motion a sense memory, the ultimate result is of course multisensory. For example, we remember not only a smell, but also the images, sounds, movements, and so on. If we consciously follow the gradual process of the construction of a sense memory, it becomes apparent how other sensory perceptions join in. Those other sensory perceptions help to make the initial vague feeling ('that smell reminds me of something') more concrete. We know much more today about this cooperation between the senses within the brain thanks to neuropsychological research on multisensory perceptions and studies of people with synaesthesia.[8] The classical concept of separate sense regions is being replaced by a concept of a brain that is sensitive to multisensory input. First, the brain wants to know about patterns of information and meanings in the world, and then tries to ascertain whether it stems from light, sound, smell, taste, touch, and so on.[9] The intersensory cooperation within the brain is far greater than was assumed until recently; I have written about this at length in my book *The Hidden Sense: Synesthesia in Art and Science*.[10]

Although sense memories are the result of cooperation between the senses, it does make a difference what kind of sensory stimulus sets the memory in motion. Psychological experiments and brain scans have shown that smells evoke more emotional memories than images and sounds, for example, whereas there is no difference in the degree of detail captured in the memory.[11] These experiments have shown that the emotional centres in the brain play a bigger role than expected in the sense memory pathways. Smell memories are the only ones for which there is a clear physiological explanation: the parts of the brain responsible for our sense of smell are situated close to the emotional centres. This leaves unanswered the question of why music, for example, is able to arouse such emotional memories: the parts of the brain responsible for our auditory perception are situated further away from the emotional centres.

How do sense memories unlock lost events from our personal past? The hippocampus plays a key role in the brain in navigating along memory pathways.[12] In sense memories, the intersensory and synaesthetic connections prompt the hippocampus to search along other memory pathways, at any rate different pathways from those used for processing verbal instructions such as remembering who you were with at a particular moment in time. The hippocampus collects sensory information for the construction or 'collage' of a memory. Thus, the uniqueness of sense memories can be partially explained by the fact that the hippocampus is led along different memory pathways than in the case of the more common verbal memories.

But how is that collage created? The collages of sensory ingredients collected by the hippocampus are presented to the prefrontal cortex, the 'decision-making centre' in the frontal lobe of the brain. The frontal lobe evidently treats the products of sense memories differently from verbally induced memories. The answers to a verbal instruction (e.g. a quiz question) are easier to assess in terms of their logic and consistency than the collages produced by sense memories. The answer to a quiz question can be assessed as right or wrong, but with a sense memory this is much more difficult.

If we imagine the working of the prefrontal cortex as being akin to that of an 'office manager' which assesses incoming information from the different 'departments' of the brain, we see that it has a difficult task. Unlike a quiz question which generates answers that are specific to the question, a smell or a melody triggers a more diffuse conglomerate of images, emotions, tactile impressions, and thoughts that is more difficult to define. Proust compared it with a work of art, the philosopher Epstein with the 'fringes of the stream of consciousness'.[13]

The prefrontal cortex is almost forced to deal with this information not like an office manager but like an artist. Not rapid decisions based on factual information, but rather a slow process in which, in a process of sifting and resifting information as when composing a collage or a symphony, the elements slowly find their place and the whole takes on a special meaning. And that is precisely the difference compared with the quiz question: instead of producing an answer that was lying waiting to be picked up by the memory, a sense memory sheds a new light on the past because the elements of the collage are arranged differently.

In evolutionary terms, sense memories are presumably an older mode of remembering—a mode that is dependent on salient characteristics of the environment and the links they have formed to memory through the previous activities of the individual, rather than on rational control and top-down search with verbal cues. Many animals seem able to record memories, but they are poor in recall. They can only retrieve their memories passively—that is, by means of involuntary cues like sensory stimulation. Voluntary retrieval is typically human.[14] It may be deduced that animals have involuntary sense memories, and perhaps a kind of Proust experience of their own.

The enchanted worlds of children

As Proust himself noted, a sense memory does not produce a factual or photographic representation of the past, but a 'coloured past'. How does that 'colouring' happen? And why do sense memories colour the past rather than presenting it in a factual and transparent way?

The answer has to be sought in the time when these experiences were stored in the brain, namely childhood. Children store important events differently from adults. Before the age of 12, the senses are important tools for a child in laying down impressions and experiences. It is only later that the child has developed sufficient language ability to record events completely in a narrative form.[15] That transition is of course a gradual one, and the differences between individual children are considerable.

The 'colouring' or 'enchantment' of early childhood memories can be explained partly by the way in which children experience and store important events. Children perceive reality in a more synaesthetic and eidetic away than adults.[16] Daniel Tammet, for example, described how as a child he perceived numbers as coloured personalities. As an adult, Tammet remembers numbers as groups of colourful individuals. His sense memories are thus coloured because his childhood memory stored them in that way. It is as if colour gives him a guide rail that leads back to his childhood.

Vladimir Nabokov did not remember the facts of his childhood, but did remember the playful sensory environments in great detail. Whether his memories were fact or fiction was of little relevance to him. He was more concerned with the physical reliving of his childhood, and produced the colourful, enchanting descriptions in his biography.

Synaesthetic and eidetic perceptions are a functional part of children's empathic development, because they help children to identify and empathize with the world around them using their sensory-physical abilities, and to understand and retain important events. They 'enchant' their world in order to grasp it and record it for later.[17]

It is striking that the adults mainly remember the enchantment of the moments in their personal past, with the factual circumstances appearing to come only in second place. The descriptions of sense memories containing numbers with personalities and family members with visualized coloured names may well have caused the reader of this book to frown on occasion; if those memories were not described in such a detailed and lifelike way, it would be easy to dismiss them as fantasy or inventions.

It could be argued that we should not believe anything of those nice children's stories. That is why it is important that those experiences are tested scientifically. If someone tells you at a party, full of conviction, that the number three is a jerk, for example, you might be inclined to laugh politely and look for someone else to talk to.[18] Fortunately, experiments have demonstrated the consistent authenticity of these and other synaesthetic and eidetic experiences with a physiological basis in the system of mirror neurons in the brain.[19] They have been found to be real empathic strategies that are used by children to understand the world around them.

This empathic approach provides an insight into how children store new impressions in their memories. The special empathic abilities of children do not disappear in later life. As adults, they help in building bridges between the present and the past, unlocking lost memories from early childhood.

The richness of sense memories in later life

A sensory stimulus can sometimes enable a 90-year-old to remember an incident that happened to them as a 4-year-old with great vividness. The timespan that can be bridged by sense memories is almost unimaginable, especially when we think how much difficulty we have in remembering a shopping list or a telephone number for just a few hours or days.

Many physical and mental functions diminish as people age, but the ability to remember the past is demonstrably stronger in people aged over 60.[20] This is not only apparent from scientific research, but can be supported by staff working in the elderly care sector, and of course by older people themselves. Older people more often adopt a reflective attitude than younger people, who are more often focused in their daily working lives on gathering information. It is therefore not surprising that older people experience more sense memories than young people. As Marcel Proust described, involuntary sense memories come in when we are not looking for them. A working environment demands continuous activity from the information-processing memory, and in a busy office environment there is less space and time to sit quietly and absent-mindedly dip a madeleine cake into a cup of tea and allow the memory to drift back to childhood. In my view, if we take our time and are aware of our surroundings, sense memories can unlock doors to lost memories in a creative process of collaborating senses that construct a collage of the past.

Pleasure and meaning

As we approach the end of our journey, some final thoughts on essential issues: Are sense memories useful? What is the point of them? How could we use them better?

Scientific research reports both cognitive and affective benefits of sense memories. The memory artists Tammet and Shereshevskii, for example, used synaesthetic sense memories to produce their exceptional cognitive memory achievements.[21] Examples of affective benefits include increased well-being of older people in residential care and nursing homes through the use of sense memories in therapies, group activities, and the configuration of the buildings, such as the use of 'snoezel' rooms and multisensory rooms.[22]

Sense memories go further back in time (to childhood episodes before the age of 12) than memories that are recovered through words, questions, and language associations. Evidently, children lay down early experiences in the form of sensory impressions, and it is only later in life, after they have developed a greater sense of language, that they lay down memories in narratives.

It is never too early to make a start with sense memories, because children have a great aptitude for sensory impression. Allow children the scope to experiment at school and at home with their senses and synaesthetic experiences. Devote attention not just to their cognitive development, but also to their physical and sensory development. Give them the space to understand objects and events with their senses, as well as in terms of language and arithmetic. Examples of this approach include the taste lessons and synaesthetic projects being carried out at primary schools.[23] Devoting attention to sensory development, for example, in creative lessons and activities, will help children to store their childhood experiences for later.

Sense memories can trigger both pleasant and unpleasant emotions. Even in people with depressive complaints for whom the past brings unpleasant associations, recalling and discussing childhood memories using photographs, smells, and other sensory stimuli can give them a more contented feeling.[24] The act of remembering itself is often enough to create a feeling of happiness. Older people in residential homes are cheered up by the sensory stimulus of memories and stories from their past, according both to reports from those concerned as well as a number of scientific studies. Good results have been achieved in reminiscence therapies with older people, using sensory stimuli to enable them to recall and re-experience memories from the past. Proust may have been the first to describe in detail how his life changed from a disconsolate, dull existence to a rich literary and artistically innovative life after he had discovered the power of sense memories and his writing career was given purpose and meaning.

Older people are very good at remembering, and sensory stimuli are a useful tool in helping them to do so. This knowledge can be applied to help older people who are not doing well. Depressed and immobile older persons can be stimulated by actively encouraging them to go outside and to engage with other older people, instead of shutting them away in their room or house. This can bring their sense memories to life, as in the examples we saw with the memory museums, Storytables, beach rooms, Smelloflowers, and cookery projects.

Exercise is good for older people, not just physically but also mentally. It generates more sensory stimuli. Older people are highly sensitive to sense memories, so by exercising and being active, their sense memories increase almost automatically, in turn increasing the their feeling of well-being. Artists and architects can help here by making the living environment more stimulating, as the BBC documentary *The Young Ones* and the Dutch documentary *Krasse Knarren* demonstrated.[25]

How could you yourself make better use of sense memories? The examples of the memory artists show how you can support and improve your memory by storing information not just abstractly as words or numbers, but also as sensory images, sounds, tastes, or smells. In fact, this is an old technique which was also used in ancient times. As well as a skill, remembering is also an art. [26] The art of remembering means visualizing information in images, though of course you could also 'audialize' it, by turning it into a song; 'tactilize' it by associating it with a touch; 'olfactorize' it by associating it with smell information; and finally, you could 'synaesthetize' the information, so that it is stored in the memory along with other, associated sensory impressions.[27]

How can you open the doors to your own museum of lost sense memories? One way is to give your senses the opportunity to work together, so that they can find the doors to the museum halls in your memory. You could also be more aware of sensory impressions in your daily environment, living more in the here and now rather than spending all your time making plans for tomorrow. You could seek out your personal history physically and mentally by revisiting places you have lived in and visited.

Let me finish by saying something about how my senses opened the doors to lost memories. I began this book with a description of a few of my sense memories, triggered by the smell of a bathroom, the texture of *Brinta* breakfast cereal, and the sound of the voice of Rod Stewart. For a long time, I paid little attention to these memories, seeing them as nothing more than amusing. It is only now that I realize that those apparently simple memories contain much more than I suspected.

Why not more often give ourselves the time and quiet space to allow ourselves to be carried back through the sense memories which spends most of its time quietly hidden in our bodies? When I came across Alain de Botton's book *How Proust Can Change Your Life* during my study of Proust, I found it contained not just amusing anecdotes about the writer's life, but also a philosophical insight into 'taking the time'. In a chapter on the therapeutic effect of remembering and well-being, de Botton writes that Proust devoted no fewer than 30 pages to the moment of falling asleep. That is an occurrence that can be dealt with in just a couple of words, as I have just done in the last sentence. As when recalling memories, however, it is not about *what* happened (I fell asleep at about midnight), but about *how* it happened and *how* it felt. That is something for which you have to take the time.

Since then, I take the time at several moments during the day. At such moments I am aware of the sensory impressions reaching my body and my sense memories are given the chance to carry me where they will. I never know in advance where I will end up. Often I am taken to places in my past that I had long since forgotten. Sometimes they appear at first sight to be meaningless, but later I often realize their significance. I then see that they had a hidden meaning, which my brain had secretly already prepared, but of which I only gradually become aware.

With a smile, I now realize that this is perhaps the answer that I was seeking to the question which runs as a common thread through this book: why do what are at first sight seemingly pointless sensory details trigger such intense emotional childhood memories? Although I am ending my book here, I hope that for you as the reader, this will not be an ending but a beginning. As soon as you have put this book down, you could make a start on being more aware of sense impressions in the present which may open doors to lost memories of your past. I wish you a great deal of pleasure on your journey!

Appendix on the neuropsychology of the memory of the senses

This appendix contains a concise neuropsychological description of the working of the memory of the senses, relating to the human memory in general, and illustrated with Proust's madeleine anecdote.

Memories

Human beings do not have just one memory, but several interrelated memory systems, for example, a short-term and long-term memory; a memory for general knowledge; a memory for automatic actions (e.g. driving a car); a memory for emotions; a memory for our personal history; and so on. Each memory follows its own pathways in the brain, and often those pathways cross each other. That is what happens, for example, when we hear an old, familiar song whilst driving and forget for a moment to change gear.[1]

Taking as a basis the Proust effect and the memory of Marcel's happy childhood which came to him after tasting a madeleine cake that he had dipped in his tea, I will distinguish between the different memory systems in order to find analytical clarity in a concise but bold classification (see Figure A.1).[2]

Remembering begins with perception. The young Marcel interpreted the taste of the madeleine cake using his knowledge, expectations, and emotions at that moment, and those impressions were stored in his memory. Later, during the moment of reminiscence as an adult, these experiences were activated by his knowledge, expectations, and emotions of *that* moment, and therefore had a different meaning from when they were first recorded in his memory. Marcel's recognition of the taste from his childhood was a manifestation of a correspondence between his current perception at that moment and his memory.

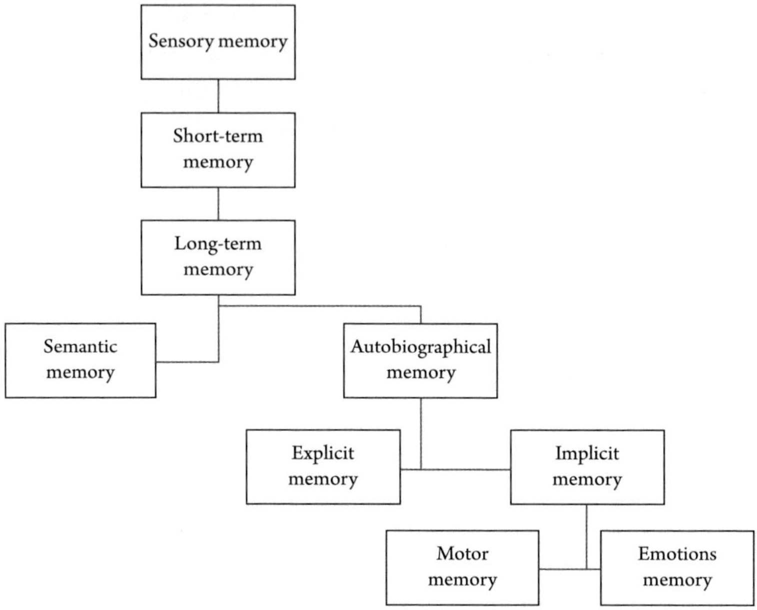

Figure A.1. Concise classification of human memory systems.

The first memory to be addressed is the sensory memory. When the young Marcel tasted the cake, the many sensory impressions he experienced at that moment were recorded in his sensory memory (not to be confused with the sense memory, e.g. the Proust effect). We are virtually unaware of the sensory memory. It retains sensory information for a very short time for processing by the *short-term memory*. It holds small amounts of information, for example, an image of the madeleine, for a short time. When the taste experience of the madeleine is given a context, either consciously or subconsciously, the information is transferred to the *long-term memory*. The long-term memory houses a collection of memories; it has been compared with a chest of drawers or the hard disk in a PC, in which all kinds of information can be stored, but in fact it is more dynamic and creative. Marcel's long-term memory recorded not only the taste and image of the cake, but also his emotions at that moment, the impressions of his aunt, her room, the morning temperature, and so on, and created new meanings to these sense impressions.

Broadly speaking, the long-term memory uses two sources: the *semantic memory* and the *autobiographical memory*. The semantic memory is used to retrieve general knowledge, for example, what day it is and where we are.[3] The autobiographical memory is where memories are stored that are related to episodes in our lives (it is therefore also called the episodic memory). When Marcel tasted another madeleine cake later in life, he automatically experienced an autobiographical memory. General knowledge, for example, the knowledge that cakes turn soft when they are dipped in hot tea, came from his semantic memory.

The autobiographical memory in turn manifests itself in an *explicit memory* and an *implicit memory*. The explicit memory holds conscious, factual information about events in the past. For example, Marcel remembered that he had entered the room and seen his aunt. The implicit memory expresses memories that are less easy to put into words, such as feelings and gestures.

The implicit memory has resources in several memories, such as an emotional and a motor memory. The *emotional memory* expresses physical feelings, moods, and other emotions. The *motor memory* is responsible for actions that we are barely aware of. Without a motor memory, for example, Marcel would have had difficulty dipping the cake in his tea without burning his fingers.

These last two memories broadly feed the domain of the *memory of the senses*. The autobiographical, implicit, emotional, and motor memories are all activated by sensory stimuli.

Parts of the brain

Several areas of the brain are involved in sense memories. Here I will discuss the main protagonists in the neural interplay which was, for example, responsible for bringing to life Marcel's flavour memory.

The brain has evolved into a structure built up of three parts (see Figure A.2). The *brain stem* (or 'reptile brain') controls vital functions such as respiration, blood circulation, waking, and sleeping. The *limbic system* is located deep within the brain and is important for emotional and behavioural functions such as caring for others, the fight or flight response, fear, and love. The biggest area of the brain is the *cerebral cortex* or *neocortex*.

The cerebral cortex is divided into two parts, the left and right hemispheres, and each hemisphere is in turn subdivided into four lobes: the *frontal lobe*, the *parietal lobe*, the *temporal lobe*, and the *occipital lobe* (see Figure A.3).

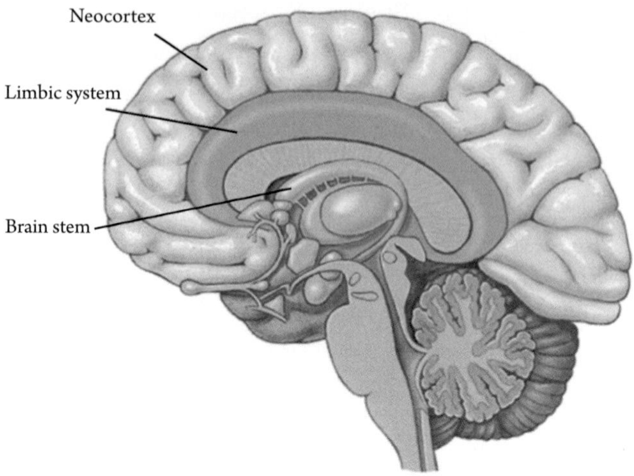

Figure A.2. Locations of the brain stem, the limbic system, and the cerebral or neocortex in the brain.

The cerebral lobes have different functions (see Figure A.4). The *frontal lobe*, situated as its name suggests at the front of the head, plays a coordinating role in gathering information, taking decisions, and directing movements. The sensory functions lie in the other lobes. The *temporal lobe* controls hearing, the *parietal lobe* the sense of touch, and the *occipital lobe* the visual functions. The sense of smell and taste are distributed across different parts of the brain, the most important of which are located in the limbic system.

Tasting the madeleine cake stimulated several of Marcel's senses, but of course his sense of taste and smell were the most important. The taste of the

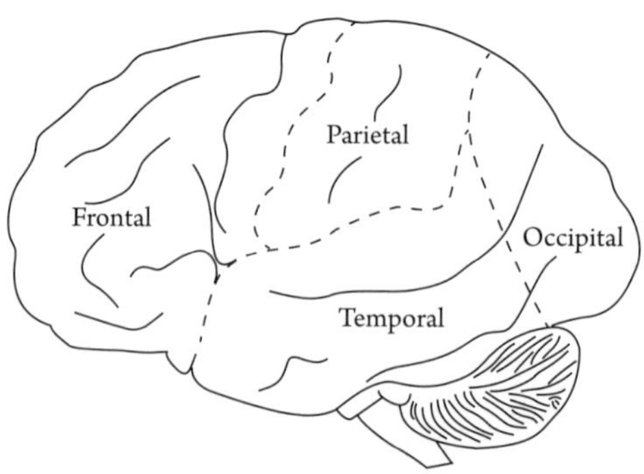

Figure A.3. The main areas of the cerebral cortex.

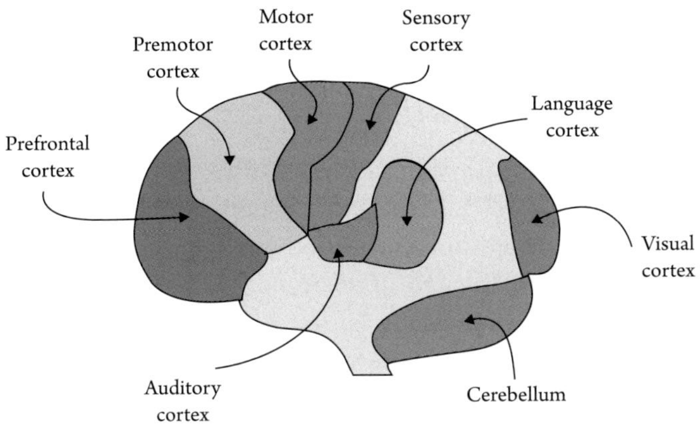

Figure A.4. Functional areas of the brain that are involved in sensory perception and sense memories.

cake stimulated taste and smell centres in the limbic system (see Chapter 6, Figure 6.1), close to the emotional centres, or *amygdalae*.

The *amygdalae* add emotional colour to events by instructing hormones to be released that are associated with the storage of impressions. They are active not only in storing those impressions, but also in remembering them.[4] The amygdalae 'underline' emotional events, as it were, in the flow of impressions, so that they can be retrieved more quickly from the memory at a later time. For Marcel, the taste of the cake triggered feelings of joy and happiness even before he explicitly remembered the time and place to which those feelings referred.

The emotions are gradually given form by the actions of the *hippocampus*, which is located close to the amygdalae and is responsible for organizing the memory. Where the amygdalae are concerned with assessing the emotional value of an event, for example, the danger of an abyss or a loving smile, the nearby hippocampus maintains an overview and lays down the circumstances in the memory. This ensures that we know later where the abyss was or what clothes the person with the loving smile was wearing at that time.

Damage to the hippocampus causes memory problems such as amnesia, a condition in which a person is unable to form new memories (like the main character in the film *Memento* by Christopher Nolan). The hippocampus is essential for our orientation, both in terms of finding our way around in the physical world and of retrieving events from the past. Among other things, damage to the hippocampus affects people's ability to navigate. It is as if the hippocampus maintains a map of incoming impressions from our surroundings and shows us in which direction we are moving. If the amygdalae warn us that we are afraid of the lion facing us, the hippocampus tells us whether we are in the jungle or the zoo.

The hippocampus brings together memory pathways running to and from the different sensory domains of the brain, which are also connected to each other. Visual impressions processed in the visual cortex at the back of the brain, auditory impressions in the temporal lobe in the central region of the brain and tactile impressions in the parietal lobe are all involved in the memory pathways and complement the smell and taste impressions. Gathering multisensory elements, the hippocampus gives rise to the 'experience' of a reconstructed past, the feeling of having one foot in the past and one in the present. Through the hippocampus, multiple sensory impressions gradually made their way into Marcel's consciousness. The room in which he was sitting at the time of the remembered event appeared, he heard the sound of his aunt's voice again, and he could once again feel the ambient temperature of that morning.

Whereas the hippocampus stores a multisensory record of the stream of consciousness, lingual memories represented in the neocortex are more concerned with generalizations or concepts derived from that stream. The initial hippocampal memory supports the development of a less detailed, schematic version in the neocortex which retains the gist of the initial memory, but not its contextual details.[5]

One drawback of the hippocampus, however, is that it only suggests possible directions and makes no distinction between present, past, and future (fantasies). Another brain function is needed to decide which of these directions is chosen. That function is performed by a small area in the frontal lobe, lying just behind the forehead. The function of this *prefrontal cortex* is akin to that of an 'office manager'. It issues search instructions, gathers information, and makes judgements as to what belongs with what and which elements or pathways are consistent with our thoughts. This enables us to decide, for example, whether the smell of new-mown grass is taking us back to an event from our childhood or into the fantasy world of a perfume advertisement we have seen. That depends on which information we need at that moment: a childhood memory or an answer to a quiz question about which advert uses the smell of new-mown grass.

The sensory impressions that are activated by the hippocampus are tested for consistency and coherence by the prefrontal cortex in the frontal lobe.

It is the prefrontal cortex, for example, which makes sure that the cup is not upside down and actually contains tea.

The prefrontal cortex and hippocampus also communicate with the *mirror neurons* in the premotor cortex and the parietal lobe. The mirror neurons provide information on physical sensations and make it possible for the adult Marcel to feel the same way he felt when he sat at the table as a small boy. The emotion was given empathic depth; Marcel felt the same joy that was felt in the body of the young Marcel.

Glossary

Amygdalae situated in the limbic system; they add emotional colour to sensory impressions and memories

Autobiographical memory contains information on episodes from our personal life

Brainstem responsible for vital functions such as heartbeat, respiration, and regulating blood pressure

Cerebral cortex the outer layer of the brain; performs sensory and cognitive functions

Emotional memory responsible for physical feelings, moods, and other emotions

Explicit memory contains the conscious part of the information in the long-term memory

Frontal lobe coordinates information-gathering, decision-making, and movements

Hippocampus located in the limbic system; gathers sensory information from the memory

Implicit memory contains the subconscious part of the information in the long-term memory

Limbic system lies deep within the brain beneath the cerebral cortex; controls emotional functions such as concern/caring, fight or fight response

Long-term memory records contextualized information for the long term

Mirror neurons located in the premotor cortex and the parietal lobe; control empathic functions relating to actions

Motor memory stores learned skills such as cycling and driving

Occipital lobe located at the rear of the cerebral cortex; responsible among other things for visual functions

Parietal lobe located in the central upper area of the cerebral cortex; responsible among other things for the sense of touch

Prefrontal cortex situated in the frontal lobe of the brain; has a coordinating and controlling function in the construction of memories

Premotor cortex located in the frontal lobe of the brain; contains functions in relation to the planning and implementation of actions

Semantic memory contains general knowledge about the world in the form of meanings, concepts, and facts

Sensory memory subconsciously retains sensory impressions for a very short period after the stimulus has ceased

Short-term memory retains a small amount of information (approximately seven units) in the conscious memory for a short time

Temporal lobe located laterally on the cerebral cortex; functions include auditory perception

V1, V2, V3, & V4 four areas located in the occipital lobe, responsible for visual functions

Notes

1. A MEMORY OF THE SENSES

1. I use the terms 'memory of the senses' and 'sense memories' with this meaning. There is also the concept of 'sensory memory', which is used in science to describe the phenomenon whereby the senses store and retain very short fragments of information. This sensory memory operates subconsciously for a very short time and is lightning fast. See the Appendix for a more detailed explanation of this and other memory terms.

2. THE PROUST EFFECT

1. The biographical information is taken from de Botton (1997), Hayman (1990), and Adé (1983).
2. Bogousslavsky (2007).
3. On Proust and dandyism, see Sillevis (2006).
4. Among others, the philosopher Alain de Botton wrote about the art of life in *How Proust can change your life*, in which he drew lessons for life from Proust's biography and oeuvre (de Botton 1997).
5. Quotation from Proust (1917–1922/2006) taken from Part 1, Combray, *Swann's Way* (*Du côté de chez Swann*).
6. Proust (1917–1922/2006), Combray, *Swann's Way*.
7. Proust (1917–1922/2006), Combray, *Swann's Way*.
8. Proust (1917–1922/2006), Combray, *Swann's Way*.
9. Proust (1917–1922/2006), Combray, *Swann's Way*.
10. Proust (1917–1922/2006), Combray, *Swann's Way*.
11. Proust (1917–1922/2006), *Time Regained* (*Le Temps retrouvé*).
12. Proust (1917–1922/2006), *Time Regained*.
13. Proust (1917–1922/2006), *Time Regained*.
14. Bergson published his principal work on the memory, *Matière et mémoire*, in 1896. See Bogousslavsky (2007) for a discussion of the role of Sollier, Proust's physician. The German philosopher and psychologist Hermann Ebbinghaus had identified three basic modes of remembering in his influential book *Über*

das Gedächtnis, in 1885: a voluntary conscious mode, an involuntary conscious mode, and a non-conscious mode.

15. Proust (1917–1922/2006), *Time Regained.*
16. Proust (1917–1922/2006), *Time Regained.*

3. THE POWER OF FRAGRANCES

1. Corbin (1988).
2. Earlier descriptions of the Proust effect can of course also be found. The Sanskrit scholar David Shulman (2006, p. 412) writes that the Sanskrit word for smell, *Vasane*, represents the *déja vu* feeling that a smell triggers in the consciousness.
3. They are often so closely interwoven in an experience that people sometimes talk about sweet smells even though nothing sweet is reaching their nostrils. The sweet smell of vanilla or strawberries is in fact a flavour perception.
4. P. Maine de Biran, *Journal* (1792–1817, part 1), cited in Corbin (1988, p. 201).
5. Cf. Schacter (1996, p. 170).
6. Maine de Biran, cited in Corbin (1988, p. 201).
7. George Sand, *Histoire de ma vie*, 1854, cited in Corbin (1988, p. 202).
8. Synaesthesia is an uncommon neurological cooperation of the senses. Synaesthetic perceptions and Proust effects have in common among other things that the cooperation of the senses generates an unusual experience which provides an insight into new aspects of reality, for example, relationships between smells and colours. I discuss synaesthesia in greater detail in Chapters 7 and 8.
9. Ch. Baudelaire (1861). *The Flowers of Evil* (*Fleurs du Mal*). Translation by William Aggeler (Fresno, CA: Academy Library Guild, 1954).
10. Baudelaire (1861). *The Flowers of Evil.* Translation by Hadi Deeb: <http://www.hadideeb.com/journal/2009/4/5/baudelaire-le-parfum.html>.
11. Corbin (1988, p. 205).
12. G. Flaubert, cited in Corbin (1988, p. 207).
13. See my discussion in van Campen (2007, pp. 106–122).
14. Proust (1917–1922/2006), *Time Regained* (*Le Temps retrouvé*),
15. Süskind (1985). The subject of odour memories is still very present in French literature, e.g. Claudel (2012).
16. For more information, see his website: <http://www.peterdecupere.net>.
17. More information can be found at <http://www.peterdecupere.com> and <http://www.skor.nl/artefact-1334-nl.html>.
18. 'Black Confetti' exhibition, Schielandshuis, Rotterdam, in 2004. See also Vogelzang (2008, pp. 6–8). In a later workshop with older people she made typical Dutch dishes and held a discussion with the participants on tastes from the past and present. See Vogelzang (2008, pp. 14–17).
19. Spence et al. (2011). See also: Crisinel and Spence (2010); Spence (2011).

20. Dinner in the dark is available in a number of cities, including Amsterdam, Paris, London, Berlin, Vienna, and New York. For Amsterdam see <http://www.artofevents.nl/themadiner_dd.html>, and in Paris at <http://paris.danslenoir.com/>.
21. *Ratatouille* is a 2007 American computer-animated comedy film produced by Pixar and directed by Brad Bird: <http://disney.go.com/disneyvideos/animatedfilms/ratatouille/main.html>.

4. LISTENING TO 'MY GENERATION'

1. Proust (1917–1922/2006), *Time Regained.*
2. Murakami (2000, pp. 6–7).
3. Holbrook and Schindler (1989).
4. Vingerhoets (2013).
5. Website: <http://top2000.radio2.nl/2006/page/verhalenarchief>. The archive is under reconstruction and will re-open on the website: <http://verhalenarchief.gahetna.nl/>.
6. Janny van Eldik-Bonouvrié (2006). *Bernardine*: see footnote 5.
7. Danny Gordijn (2006). *Noem me Sebastian* ['Call me Sebastian']. See footnote 5.
8. Herman (2006). I see her cycling past my window when I hear Tom Jones: see footnote 5.
9. Douwe Gerlof Heeringa (2006). The smell of shower gel and the sounds of 'Pride': see footnote 5.
10. See also Van Dijck (2009).
11. Peesapati et al. (2010). *Pensieve* site at Cornell University: <http://pensieve.cornellhci.org/>.
12. Peesapati et al. (2010).
13. Peesapati et al. (2010).
14. Cf. Habermas (2012).

5. THE ART OF MEMORY

1. Yates (1992, p. 102ff.)
2. Yates (1992, p. 106ff.)
3. Gombrich (1960).
4. See also Wainwright (2004).
5. See, e.g. Pastore (1971).
6. Cited in *de Volkskrant* newspaper in a review of a Dali exhibition in Tate Modern, London, 29 June 2007.
7. McLuhan (1964).
8. McLuhan (2005, pp. 43–52).
9. Website: <http://www.universalstudiosentertainment.com/eternal-sunshine-of-the-spotless-mind>.

10. The technique does not exist. The makers of the film created a spoof website on which the fictitious technique used by Lacuna Inc. was explained. The website, which no longer exists, contained a quasi-scientific description of the inventor of the technique and the procedures used.
11. See Van Dijck (2007); Bijsterveld and Van Dijck (2009).
12. A number of scientific research projects have been launched, such as the Project Sound Souvenirs Maastricht. Website: <http://www.soundsouvenirs.org/en/>.
13. See Van Dijck (2007, pp. 153–169).
14. Cf. the use of new digital devices like SenseCam in memory studies. SenseCam (also known as ViconRevue) is a small, wearable digital camera fitted with electronic sensors (e.g. to detect light, heat) that can automatically trigger thousands of photographs in a single day (St Jacques 2012).

6. THE HIPPOCAMPUS OF PROUST

1. One science journalist actually regarded Proust as a forerunner of modern neuroscientific research into the memory (Lehrer 2007). Curious detail: Marcel Proust came from a family of doctors and was friends with many neurologists (Bogousslavsky 2007).
2. Also called the Proust experience or Proust phenomenon. In essence, all terms refer to the anecdote of the taste of the madeleine cake dipped in tea (see Chapter 2).
3. Laird (1935).
4. Laird (1935, p. 128).
5. Draaisma (2004, 2008).
6. Personal communication, Douwe Draaisma (26 May 2008).
7. Personal communication, Douwe Draaisma (26 May 2008).
8. The scientific article mentions the following smells: coffee, Johnson & Johnson baby powder, cinnamon, cigarettes, rubbing alcohol, mint, mothball, Ivory soap, banana, onion, peanut butter, chocolate, band-aids, bourbon, popcorn.
9. See Chapter 2 for an explanation of the voluntary and involuntary memory.
10. Chu and Downes (2000).
11. This phenomenon is known as the 'reminiscence bump' (Rubin et al. 1986).
12. Berntsen and Rubin (2002); Schlagman et al. (2007).
13. Berntsen (2009, 2012). Incidentally, on average involuntary memories occurred three times more frequently than voluntary memories.
14. Cf. Brewin et al. (2010).
15. Chu and Downes (2000).
16. Bohn and Berntsen (2008); Habermas and de Silveira (2008); Habermas et al. (2009).
17. Habermas et al. (2010); Van Abbema and Bauer (2005).
18. O'Kearney et al. (2007); Reese et al. (2010).

19. Bauer et al. (2005); Habermas et al. (2010).
20. Bohn and Berntsen (2008); Bruce et al. (2005); Habermas and Bluck (2000).
21. See Bauer (2007) for a review.
22. Bauer (2012).
23. See, e.g. review by Bauer (2012).
24. Herz and Cupchik (1992).
25. Herz and Schooler (2002); Herz et al. (1998, 2004).
26. Herz (1996).
27. See Chapter 2.
28. LeDoux (1996, 2007).
29. LeDoux (1996, 2007).
30. D'Argembeau (2012); Moscovitch (2012); St. Jacques (2012).
31. Gottfried et al. (2004).
32. The researchers taught their subjects to link smells to photographs, for example, a photograph of a duck and the smell of roses. When the subjects viewed the 'smell photographs', both the hippocampus and the olfactory centres became active, whereas when looking at other photographs only the hippocampus was activated (Gottfried et al. 2004).
33. Gottfried et al. (2004).
34. Herz (1996, 1998); Herz and Schooler (2002).
35. See also illustrations in the Appendix.
36. Epstein (2004).
37. Bartsch (2005, p. 96ff.).
38. Cf. Schacter (1996).
39. Bartsch terms this subconscious process 'concept formation', i.e. the formation of new forms from the elements present (Bartsch 2005, p. 96ff.).
40. Rubin and Greenberg (2003), Schacter (1996), Rubin and Berntsen (2012).
41. Proust (1917–1922/2006), *Time Regained.*
42. Maguire et al. (2000).
43. Henneman et al. (2009).
44. Marks (2012) distinguishes three stages in the Proust effect: (1) a break in the flow of activity (a smell disrupts the activity) continuity of experience, (2) emotion (is felt), (3) source of memory (resolved).

7. NABOKOV AS A TODDLER IN ST PETERSBURG

1. Nabokov (2002).
2. Bauer (2012).
3. Nabokov (1989/1951, chapter 1).
4. Nabokov (1989/1951, chapter 1).
5. Nabokov (1989/1951, chapter 1).
6. Nabokov (1989/1951, chapter 1).

7. Schacter (1996) distinguishes between the birds eye perspective and the perspective perceived in memories.
8. Nabokov (1989/1951, chapter 1).
9. Cf. van Campen (2007); Simner and Hubbard (2013).
10. Nabokov (1989/1951, chapter 1).
11. Dann (1998).
12. Jaensch (1930); Werner (1948); Van Campen (2007).
13. See, e.g. the accounts of synaesthetes in Dittmar (2007) and Sagiv et al. (2011), and artistic experiments with synaesthetic perceptions in Evers (2012) and Daniels and Naumann (2010).
14. Lev (2006).
15. See Pajo (2006) for a discussion of the different explanations of Mozart's use of coloured scores.
16. Purdy (1936), quoted in Dann (1998, pp. 106–107).
17. Cytowic (2002, p. 195).

8. THE LITTLE BRICOLEUR

1. De Groot (1946).
2. Jaensch (1930).
3. Werner (1948).
4. Cf. Iacoboni (2008, p. 93); Cytowic and Eagleman (2009, pp. 105–108).
5. See, e.g. the personal reports of synaesthetes with eidetic memories in Dittmar (2007).
6. Werner (1948) used the term 'syncretic perception' for the initial stage of as yet undifferentiated new impressions when observing objects such as tables and chairs, and 'discrete perception' for the provisional end stage.
7. Werner (1948, p. 53ff.).
8. Stumpel and Kieft (1989).
9. See, e.g. Stumpel and Kieft (1989) and Gombrich (1960).
10. Werner (1948, p. 71) quotes from Kandinsky's autobiography: (Kandinsky, '1901-1913', Der Sturm, 1913, p. 5, 6, 17).
11. Quotation taken from Smilek et al. (2007).
12. Simner and Holenstein (2007).
13. Simner et al. (2006).
14. Quotation taken from Simner and Holenstein (2007).
15. There were also some participants in the experiment who felt a touch on their own right cheek when the other person was touched on the right cheek (Banissy and Ward 2007). Approximately 1.4% of people experience these sensations to such a marked degree as this (Banissy et al. 2009).
16. Banissy and Ward (2007); Banissy et al. (2009). Earlier brain research had also identified essential differences (Blakemore et al. 2005).
17. The mirror touch synaesthetes scored higher on all empathy scales. The difference was only significant on the 'emotional response' scale. The number of

participants in the study was however low (10 versus 20 persons), so that the chance of finding significant differences was small (Banissy and Ward 2007).

18. Banissy and Ward (2007).
19. Ramachandran (2000).
20. Jabbi et al. (2007). The researchers used the Interpersonal Reactivity Scale, which distinguishes different aspects or dimensions: empathic concern, fantasy, perspective-taking.
21. Gazzola et al. (2006).
22. Van der Gaag et al. (2007).
23. The researchers link recent ideas on mirror neurons to an early psychological theory of empathic perceptions by Theodor Lipps (1907), a contemporary of Heinz Werner and Jaensch. Lipps posited that sensory figurations (images, music) are felt with the *Gestalt* of the observer's own subjectively felt body (Van der Gaag et al. 2007, p. 181). See also Iacoboni (2008) on Lipps (1907) and Merleau-Ponty (1945).
24. Cf. my earlier comments on Abstract Expressionism in painting (Kandinsky) and animation (Disney).
25. Castelli et al. (2000).
26. Smilek et al. (2007).
27. See Keysers (2011) for more discussion of the social and ethical implications of research into mirror neurons.
28. Emrich (in preparation).

9. DO SENSE MEMORIES MAKE YOU HAPPIER?

1. For more information, see the website of CIAN: Centre for Aromatherapy: <http://www.cian-be.com/>.
2. Peru Balsam is extracted from the Myroxylon balsamum pereirae, a tree from Central America. Peru balsam has uses in medicine, pharmaceuticals, in the food industry and in perfumery. It has been used as a cough suppressant,
3. Quotation by Veerle Waterschoot. Interview held on 26 April 2008 in Haasdonk, Belgium.
4. Quotation of Veerle Waterschoot. Interview held on 26 April 2008 in Haasdonk, Belgium.
5. Quotation of Veerle Waterschoot. Interview held on 26 April 2008 in Haasdonk, Belgium.
6. Damian (1995); Damian and Damian (2006).
7. More information on Prik! may be found on the RASA website: <http://www.rasa.be>.
8. See website <http://www.smaaklessen.nl> and Wind (2006) and the international site Sapere: <http://sapere.ebaia.com/en/taste-classes.html>.
9. In Britain, the chef Jamie Oliver started a familiar campaign 'Feed me better' in schools. Cf. <http://en.wikipedia.org/wiki/Jamie's_School_Dinners>.
10. Simner and Ward (2006); Ward and Simner (2003); Ward et al. (2005).

11. Quotation by James Wannerton taken from the BBC documentary *Derek Tastes of Ear Wax*. Website: <http://news.bbc.co.uk/2/hi/uk_news/scotland/edinburgh_and_east/7320452.stm>.
12. See Chapters 7 and 8 for discussions of the phenomenon of synesthesia.
13. Simner and Ward (2006); Ward and Simner (2003); Ward et al. (2005).
14. Lambrechts (2003).
15. Lambrechts (2003).
16. Lambrechts (2003, p. 9).
17. Website: <http://www.geheugenvannederland.nl/?/en/homepage>.
18. See, e.g. the Dutch website <http://nl.wikipedia.org/wiki/Beelddenken> and the international website <http://en.wikipedia.org/wiki/Visual_thinking>.
19. Simner et al. (2009).

10. UPLIFTING MUSICAL MEMORIES

1. Oliver Sacks, Interview in *de Volkskrant* newspaper, 13 October 2007.
2. Sacks (2007).
3. Sacks (2007).
4. Cf. Chapter 7.
5. Email correspondence with Marcia Smilack, 7 April 2009.
6. Marcia Smilack, weblog at <http://www.marciasmilack.com>.
7. Marcia Smilack, Artist Statement at <http://www.marciasmilack.com>.
8. Email correspondence with Marcia Smilack, 10 April 2009.
9. Seaberg (2011).
10. Seaberg (2011, pp. 98–99).
11. Styron (1989).
12. Styron (2003/1989), cited in Sacks (2007).
13. Email correspondence with Marcia Smilack, 7 April 2009. Rose is Rose Styron-Burgunder, widow of William Styron.
14. Cited in Sacks (2007, p. 345).
15. More information is available on the Music and Memory website: <http://musicandmemory.org/>.
16. More information is available on the Music for Life project website: <http://www.newdynamics.group.shef.ac.uk/music-for-life.html>.
17. See Becker (2008) and Moscoviter (2006). More information may be found on the website <www.stichtinghumanitas.nl>.
18. More information may be found on the website <www.verhalentafel.nl>.
19. *Brabants Dagblad* newspaper, 7 November 2007. M-book was developed by Mariet Schreurs and Helma van Rijn at the Faculty of Industrial Design at Delft University of Technology.
20. Sharot et al. (2007); Sharot (2011).
21. D'Argembeau and Van der Linden (2008).
22. Bryant et al. (2005).

23. Bryant et al. (2005, pp. 236–237).
24. Bryant et al. (2005, p. 240).
25. Bryant et al. (2005, p. 251). Comparable studies have also been carried out on older people by Westerhof et al. (2010).
26. Steunenberg and Bohlmeijer (2010).
27. Quotation in Steunenberg and Bohlmeijer (2010).
28. Berntsen (2009, 2012)
29. D'Argembeau et al. (2003); Williams (1996).
30. Pot et al. (2010).
31. See also Hamilton (2010) for a discussion of the role of the Proust effect in oral histories from a cultural anthropological perspective.
32. Woods et al. (2005).
33. Bohlmeijer et al. (2011).
34. Hulsegge and Verheul (1987).
35. Van Weert (2004).
36. Website: <http://www.nursing.nl/home/nieuw/6987/dementerenden-zonnen-in-strandkamer>.
37. The video can be viewed online at: <http://www.rtvnh.nl/programma/34/Van+Zuks+Dus>.

11. REMEMBERING 20,000 DIGITS OF PI

1. See Chapters 7 and 8 for definitions and theories of eidetic and synaesthetic perceptions.
2. Cf. the classical learning curve (Ebbinghaus 1885).
3. Luria (1968).
4. Luria (1968).
5. Luria (1968).
6. Luria (1968).
7. See also the discussion of work by the psychologist Werner and the artist Disney in Chapter 8.
8. In adults, the age of the earliest memory is typically around 3 to 3½ years (36–42 months), though there are pronounced individual differences (Bauer 2012).
9. Luria (1968, p. 77).
10. Luria (1968)., pp. 77–78).
11. 'Pi memory feat.' Oxford University. 15 March 2004. Retrieved 15 March 2013. <http://www.ox.ac.uk/media/news_stories/2004/040315_1.html>.
12. Tammet (2009, p. 58).
13. Tammet (2012, p. 116).
14. Tammet (2012, p. 117).
15. Compare the descriptions of numbers and personalities in Chapter 8.
16. Tammet (2007, p. 2).
17. Tammet (2007, pp. 4–5).

18. Tammet (2007, p. 32).
19. Tammet (2007, pp. 86–87).
20. Compare the studies by Smilek et al. (2002) and Bergfeld Mills et al. (2006).
21. Azoulai et al. (2005).
22. Bor et al. (2007).
23. Smilek et al. (2002).
24. Bergfeld Mills et al. (2006).
25. Yaro and Ward (2007).
26. Yaro and Ward (2007).
27. Rothen and Meier (2009, 2010).
28. See also the following studies on memory functions among groups of synaesthetes: Gross et al. (2011); Radvansky et al. (2011).
29. Meier and Rothen (2013).
30. Dittmar (2009).
31. Dittmar (2007, pp. 122, 166).
32. See, e.g. Duffy (2001); Seaberg (2011).
33. The hippocampus plays a central role in memorizing and navigation (see Appendix).
34. This phenomenon was described at the end of the nineteenth century by the German psychologist Eugen Bleuler (Dittmar 2007, p. 156ff.).
35. Ward (2008, p. 137).
36. Cf. van Campen and Froger (in preparation).

12. HOW PEOPLE COLOUR THEIR PAST

1. The advertisement in the newsletter *Synesthesie Nederland* brought responses from 21 people. They all completed a written questionnaire. Based on the responses, eight were invited for a verbal interview. Parts of five of these interviews have been used in this book. The interviews were conducted by the author in the spring of 2008.
2. Interview with Janneke van Dijk, 14 March 2008.
3. Interview with Janneke van Dijk, 14 March 2008.
4. Interview with Janneke van Dijk, 14 March 2008.
5. Interview with Janneke van Dijk, 14 March 2008.
6. Interview with Janneke van Dijk, 14 March 2008.
7. Interview with Janneke van Dijk, 14 March 2008.
8. Interview with Elsbeth Rabus, 4 April 2008.
9. Interview with Elsbeth Rabus, 4 April 2008.
10. Interview with Elsbeth Rabus, 4 April 2008.
11. Interview with Elsbeth Rabus, 4 April 2008.
12. Interview with Elsbeth Rabus, 4 April 2008.
13. Van Campen (2007, pp. 38–42); van Campen and Froger (in preparation).
14. Interview with Clara Froger, 4 March 2008.
15. Interview with Clara Froger, 4 March 2008.

16. Interview with Clara Froger, 4 March 2008.
17. Interview with Clara Froger, 4 March 2008.
18. Interview with Clara Froger, 4 March 2008.
19. Interview with Nelleke van den Oever, 21 March 2008.
20. Interview with Nelleke van den Oever, 21 March 2008.
21. Interview with Nelleke van den Oever, 21 March 2008.
22. Interview with Nelleke van den Oever, 21 March 2008.
23. Interview with Nelleke van den Oever, 21 March 2008.
24. Interview with Martina Strusny, 18 March 2008.
25. Interview with Martina Strusny, 18 March 2008.
26. Interview with Martina Strusny, 18 March 2008.
27. Interview with Martina Strusny, 18 March 2008.
28. Interview with Martina Strusny, 18 March 2008.
29. See more personal accounts by synaesthetes in Duffy (2001); Seaberg (2011).
30. Compare Nabokov's sense memories of his early childhood in Chapter 7.
31. Habermas (2012).

13. ENJOYING SENSE MEMORIES

1. See, e.g. the project 'Tijd van je leven' ('Time of your life') (<http://www.detijdvanjeleven.nl>) in the Netherlands and *Music for Life* the UK (<http://www.newdynamics.group.shef.ac.uk/music-for-life.html>).
2. Cf. Appendix and Glossary.
3. Schacter (1996).
4. For more examples of artistic projects involving verbal and sense memories, see Verbeek and van Campen (2013).
5. Brewin et al. (2010). Others have suggested that involuntary and voluntary memories follow the same procedures and have the same access to the past (Berntsen 2012).
6. Neurological researchers explain this through the proximity of the olfactory centres and emotional centres in the brain (cf. experiments by Herz et al. in Chapter 6).
7. Hofer (1934).
8. Cf. Calvert et al. (2004); Simner and Hubbard (2013).
9. Rosenblum (2010)
10. Van Campen (2007).
11. See the experiments by Herz et al. discussed in Chapter 6.
12. See the experiment by Gottfried et al discussed in Chapter 6.
13. Proust (1917–1922/2006); Epstein (2004); and see Chapter 6.
14. Berntsen (2009, 2012); Donald (1991, 2012).
15. Chu and Downes (2000); Habermas (2012).
16. Unfortunately, there is little empirical evidence for this, with the exception of a few historical studies by Jaensch and Werner discussed in Chapter 8.

17. In the preceding chapters (e.g. 7, 8, 11, and 12), we have seen examples of the sensory-empathic perceptions of children coming out later in their memories as adults who describe those impressions as coloured and almost enchanting experiences.
18. See Chapter 8.
19. See discussion of mirror neurons in Chapter 8.
20. Kenyon et al. (2010).
21. See Chapter 11.
22. See Chapter 9.
23. See the discussion of the taste lessons (Smaaklessen) and Prik! projects in Chapter 9.
24. See Chapter 9.
25. See the BBC TV documentary *The Young Ones* (<http://www.bbc.co.uk/programmes/b00tq4d3>) in the UK and the Dutch remake in the form of the documentary *Krasse Knarren* (<http://www.omroepmax.nl/krasseknarren>).
26. See the discussion of the study by Frances in Chapter 5.
27. As Daniel Tammet learned to do with numbers as a child (cf. Chapter 11).

APPENDIX ON THE NEUROPSYCHOLOGY OF THE MEMORY OF THE SENSES

1. The idea that people have different memories is much older, however. The memory psychologist Daniel Schacter (1996), who widely publicized and researched the notion of different memories, refers to the French philosopher Pierre Maine de Biran, who wrote as long ago as the beginning of the nineteenth century that the memory can be subdivided into three systems: a memory for ideas, a memory for feelings, and a memory for habits. Almost a century later, Proust's mentor Henri Bergson distinguished the conscious memory of ideas from the subconscious memory of habits (Bogousslavsky 2007).
2. In scientific research, the division between well-known memory systems such as short-term memory and long-term memory is still an issue for debate (cf. Baddeley 2012; Rubin 2012).
3. Damage to the semantic memory has a major impact on autobiographical memories. We see this in people with dementia, who have difficulty in assigning meaning to impressions (Where am I? What day is it? Who is this person?) and therefore in remembering recent events.
4. Human brain scans show that the amygdalae, the hippocampus, and the prefrontal cortex are more active in autobiographical memories than when remembering names, for example (Greenberg et al. 2005).
5. Winocur et al. (2010).

Bibliography

Adé, G. (1983). Marcel Proust. In: Bachrach, A. and Assche, A. van (eds.), *Moderne encyclopedie van de wereldliteratuur*, pp. 355–357. Haarlem: De Haan.

Azoulai, S., Hubbard, E., and Ramachandran, V. S. (2005). Does synesthesia contribute to mathematical savant skills? *Journal of Cognitive Neuroscience*, Supplement: 69.

Baddeley, A. (2012). Reflections on autobiographical memory. In: Berntsen, D. and Rubin, D. (eds.), *Understanding Autobiographical* Memory, pp. 70–88. Cambridge: Cambridge University Press.

Banissy, M.J., Cohen Kadosh, R., Maus, G.W., Walsh, V., and Ward, J. (2009). Prevalence, characteristics and a neurocognitive model of mirror-touch synaesthesia. *Experimental Brain Research*, 198:261–272, doi: 10.1007/s00221-009-1810-9.

Banissy, M.J. and Ward, J. (2007). Mirror-touch synesthesia is linked with empathy. *Nature Neuroscience*, 10:815–816.

Baron-Cohen, S. and Harrison, J.E. (eds.) (1997). *Synaesthesia: Classic and Contemporary Reading*. Cambridge: Blackwell.

Bartsch, R. (2005). *Memory and Understanding. Concept formation in Proust's A la recherche du temps perdu*. Amsterdam: John Benjamins.

Bauer, P.J. (2012). The life I once remembered: the waxing and waning of early memories. In: Berntsen, D. and Rubin, D. (eds.), *Understanding Autobiographical Memory*, pp. 205–225. Cambridge: Cambridge University Press.

Bauer, P.J., Stark, E.N., Lukowski, A.F., Rademacher, J., Van Abbema, D.L., and Ackil, J.K. (2005). Working together to make sense of the past: mothers' and children's use of internal states language in conversations about traumatic and non-traumatic events. *Journal of Cognition and Development*, 6:463–488.

Becker, H. (2008). *Verboden af te blijven! Het herinneringsmuseum, spiegel van de Humanitas zorgfilosofie* [No permission to not touch! Memory museum, mirror of the Humanitas philosophy of care]. Delft: Eburon.

Bergfeld Mills, C., Innis, J., Westendorf, T., Owsianiecki, L., and McDonald, A. (2006). Effect of a synesthete's photisms on name recall. *Cortex*, 42:155–163.

Berntsen, D. (2009). *Involuntary Autobiographical Memories: An Introduction to the Unbidden Past*. Cambridge: Cambridge University Press.

Berntsen, D. (2012). Spontaneous recollections: involuntary autobiographical memories are a basic mode of remembering. In: Berntsen, D. and Rubin, D. (eds.),

Understanding Autobiographical Memory, pp. 290–310. Cambridge: Cambridge University Press.

Berntsen, D. and Rubin, D.C. (2002). Emotionally charged autobiographical memories across the lifespan: the recall of happy, sad, traumatic, and involuntary memories. *Psychology and Aging*, 17:636–652.

Bijsterveld, K. and Dijck, J. van (2009). *Sound Souvenirs: Audio Technologies, Memory and Cultural Practices*. Amsterdam: Amsterdam University Press.

Blakemore, S.J., Bristow, D., Bird, G., Frith, C., and Ward, J. (2005). Somatosensory activations during the observation of touch and a case of vision-touch synaesthesia. *Brain*, 128:1571–1583.

Bogousslavsky, J. (2007). Marcel Proust's diseases and doctors: the neurological story of a life. *Frontiers of Neurology and Neuroscience*, 22:89–104.

Bohlmeijer, E.T., Steunenberg, B., and Westerhof, G.J. (2011). Reminiscence and mental health, empirical underpinning of interventions. *Tijdschrift voor Gerontologie en Geriatrie*, 42:7–16.

Bohn, A. and Berntsen, D. (2008). Life story development in childhood: the development of life story abilities and the acquisition of cultural life scripts from late middle childhood and adolescence. *Developmental Psychology*, 44:1135–1147.

Bor, D., Billington, J., and Baron-Cohen, S. (2007). Savant memory for digits in a case of synaesthesia and asperger syndrome is related to hyperactivity in the lateral prefrontal cortex. *Neurocase*, 13:311–319.

Botton, A. de (1997). *How Proust Can Change Your Life*. London: Picador.

Breeuwsma, G. (1993). *Alles over ontwikkeling: Over de grondslagen van de ontwikkelingspsychologie* [All on Development: The Foundations of Developmental Psychology]. Amsterdam: Boom.

Brewin, C.R., Gregory, J.D., Lipton, M., and Burgess, N. (2010). Intrusive images and memories in psychological disorders: characteristics, neural basis, and treatment implications. *Psychological Review*, 117:210–232.

Bruce, D., Wilcox-O'Hearn, A., Robinson, J.A., Phillips-Grant, K., Francis, L., and Smith, M.C. (2005). Fragment memories mark the end of childhood amnesia. *Memory & Cognition*, 33:567–576.

Bryant, F., Smart, C.M., and King, S.P. (2005). Using the past to enhance the present: Boosting happiness through positive reminiscence. *Journal of Happiness Studies*, 6:227–260.

Calvert, E., Spence, C., and Stein, B.E. (eds.) (2004). *Handbook of Multisensory Perception*. Cambridge MA: MIT Press.

Campen, C. van and C. Froger (in preparation). How children paint music: Development and awareness of the collaboration of the senses in children through art education projects. In: Söffing, C. and Sinha, J. (eds.), *Synaesthesia and Children: Creativity and Learning*. Luxemburg: Verlag Synaisthesis.

Campen, C. van (2007). *The Hidden Sense: Synesthesia in Art and Science*. Cambridge, MA: MIT Press.

Campen, C. van (1994). *Beeldillusies. Gezichtsbedrog in afbeeldingen sinds de renaissance* [Pictorial Illusions: Trompe l'Oeil in Pictures since the Renaissance]. De Bilt: Cantecleer.

Castelli, F., Happe, F., Frith, U., and Frith, C. (2000). Movement and mind: a functional imaging study of perception and interpretation of complex intentional movement patterns. *Neuroimage*, 12:314–325.

Chu, S. and Downes, J.J. (2000). Long live Proust: the odour-cued autobiographical memory bump. *Cognition*, 75:B41–B50.

Claudel, P. (2012). *Parfums*. Paris: Éditions Stock.

Corbin, A. (1988). *The Foul and the Fragrant: Odor and the French Social Imagination*. Cambridge, MA: Harvard University Press.

Crisinel, A.S. and Spence, C. (2010). A sweet sound? Food names reveal implicit associations between taste and pitch. *Perception* 39:417–425.

Cytowic, R.E. (1986). *The Man Who Tasted Shapes*. Cambridge: Bradford.

Cytowic, R.E. (2002). *Synesthesia: A Union of the Senses* (2nd edn). Cambridge, MA: MIT Press.

Cytowic, R.E. and Eagleman, D.M. (2009). *Wednesday is Indigo Blue: Discovering the Brain of Synesthesia*. Cambridge, MA: MIT Press.

Damian, P. (1995). *Aromatherapy. Scents and Psyche. Using Essential Oils for Physical and Emotional Well-Being.* Rochester, VT: Healing Arts Press.

Damian, P. and Damian, K. (2006) Environmental fragrancing. In: Drobnick, J. (ed.), *The Smell Culture Reader*, pp. 148–160. Oxford: Berg.

Daniels, D. and Naumann, S. (2010). *See this Sound. Audiovisuology Compendium*. Köln: Walther König.

Dann, K. (1998). *Bright Colors Falsely Seen: Synaesthesia and the Search for Transcendental Knowledge*. New Haven, CT: Yale University Press.

D'Argembeau, A. (2012). Autobiographical memory and future thinking. In: Berntsen, D. and Rubin, D. (eds.), *Understanding Autobiographical Memory*, pp. 311–330. Cambridge: Cambridge University Press.

D'Argembeau, A., Comblain, C., and Van der Linden, M. (2003). Phenomenal characteristics of autobiographical memories for positive, negative, and neutral events. *Applied Cognitive Psychology*, 17:281–294.

D'Argembeau, A. and Van der Linden, M. (2008). Remembering pride and shame: Self-enhancement and the phenomenology of autobiographical memory. *Memory*, 16:538–547.

Dijck, J. van (2007). *Mediated Memories in the Digital Age*. Stanford, CA: Stanford University Press.

Dijck, J. van (2009). Remembering songs through telling stories: pop music as a resource for memory. In: Bijsterveld, K. and Dijck, J. van (eds.), *Sound Souvenirs: Audio Technologies, Memory and Cultural Practices*, pp. 107–121. Amsterdam: Amsterdam University Press.

Dittmar, A. (ed.) (2007/2009). *Synästhesien. Roter Faden durchs Leben? / Synaesthesia: A 'Golden Thread' Through Life?* Essen: Verlag Die Blaue Eule.

Donald, M. (1991). *Origins of the Modern Mind: Three Stages in the Evolution of Culture and Cognition*. Cambridge, MA: Harvard University Press.

Donald, M. (2012). Evolutionary origins of autobiographical memory: a retrieval hypothesis. In: Berntsen, D. and Rubin, D. (eds.), *Understanding Autobiographical Memory*, pp. 290–310. Cambridge University Press.

Draaisma, D. (2001). *Metaphors of Memory: A History of Ideas about the Mind*. Cambridge: Cambridge University Press.

Draaisma, D. (2004). *Why Life Speeds Up As You Get Older: How Memory Shapes our Past*. Cambridge: Cambridge University Press.

Draaisma, D. (2008). *De heimweefabriek. Geheugen, tijd & ouderdom* [The Nostalgia Factory: Memory, Time & Ageing]. Groningen: Historische Uitgeverij.

Draaisma, D. (2010). *Vergeetboek* [Forget-book]. Groningen: Historische Uitgeverij.

Duffy, P.L. (2001). *Blue Cats and Chartreuse Kittens. How Synesthetes Color Their Worlds*. New York: W. H. Freeman.

Ebbinghaus, H. (1885). *Über das Gedächtnis. Untersuchungen zur experimentellen Psychologie* [On Memory: Studies into Experimental Psychology]. Leipzig: Verlag von Duncker & Humblot.

Emrich, H.M. (in preparation). Crucial questions as to the origin of synaesthesia during childhood. In: Söffing, C. and Sinha, J. (eds.) *Synaesthesia and Children: Creativity and Learning*. Luxemburg: Verlag Synaisthesis.

Epstein, R. (2004). Conciousness, art and the brain: lessons from Marcel Proust. *Consciousness and Cognition*, 13:213–240.

Evers, F. (2012). *The Academy of the Senses. Synesthetics in Science, Art and Education*. The Hague: Artscience Interfaculty Press.

Gaag, C. van der, Minderaa, R., and Keysers. C. (2007). Facial expressions: what the mirror neuron system can and cannot tell us. *Social Neuroscience*, 2:179–222.

Gazzola, V., Aziz-Zadeh, L., and Keysers, C. (2006). Empathy and the somatotopic auditory mirror system in humans. *Current Biology*, 16:1824–1829.

Gombrich, E.H. (1960). *Art & Illusion: A Study in the Psychology of Pictorial Representation*. Oxford: Phaidon.

Gottfried, J.A., Smith, A.P.R., Rugg, M.D., and Dolan, R.J. (2004). Remembrance of odors past: Human olfactory cortex in cross-modal recognition memory. *Neuron*, 42:687–695.

Greenberg, D.L., Rice, H.J., Cooper, J.J., Cabeza, R., Rubin, D.C, and La-Bar, K.S. (2005). Co-activation of the amygdala, hippocampus and inferior frontal gyrus during autobiographical memory retrieval. *Neuropsychologia*, 43:659–674.

Groot, A.D. de (1946). *Het denken van den schaker: argumenten voor een nieuwe traditie* [The Thinking Process of the Chess Player: Arguments for a New Tradition]. Amsterdam: Noord-Hollandsch Uitgevers Mij.

Gross, V.C., Neargarder, S., Caldwell-Harris, C.L., and Cronin-Golomb, A. (2011). Superior encoding enhances recall in color-graphemic synesthesia. *Perception*, 40:196–208.

Haber, R.N. and Haber, R.B. (1964). Eidetic imagery. I. Frequency. *Perceptual and Motor Skills*, 19:131–138.

Habermas, T. (2012). Identity, emotion, and the social matrix of autobiographical memory: a psychoanalytic narrative view. In: Berntsen, D. and Rubin, D. (eds.), *Understanding Autobiographical Memory*, pp. 33–53. Cambridge: Cambridge University Press.

Habermas, T. and Bluck, S. (2000). Getting a life: the emergence of the life story in adolescence. *Psychological Bulletin*, 126:748–769.

Habermas, T. and de Silveira, C. (2008). The development of global coherence in life narratives across adolescence: temporal, causal, and thematic aspects. *Developmental Psychology*, 44:707–721.

Habermas, T., Ehlert-Lerche, S., and de Silveira, C. (2009). The development of the temporal macrostructure of life narratives across adolescence: beginnings, linear narrative form, and endings. *Journal of Personality*, 77:527–560.

Habermas, T., Negele, A., and Mayer, F.B. (2010). 'Honey, you're jumping about': mothers' scaffolding of their children's and adolescents' life narration. *Cognitive Development*, 25:339–351.

Hamilton, P. (2010). The Proust effect: oral history and the senses. In: Ritchie, D.A. (ed.), *The Oxford Handbook of Oral History*, pp. 219–232. Oxford: Oxford University Press.

Harrison, J. (2001). *Synaesthesia: The Strangest Thing*. Oxford: Oxford University Press.

Haverkamp, M. (2013). *Synesthetic Design: Handbook for a Multisensory Approach*. Basel: Birkhäuser.

Hayman, R. (1990). *Proust: A Biography*. London: Minerva.

Henneman, W.J.P., Sluimer, J.D., Barnes, J., Flier, W.M. van der, Sluimer, I.C., Fox, N.C., Scheltens, P., Vrenken, H., and Barkhof, F. (2009). Hippocampal atrophy rates in Alzheimer disease: added value over whole brain volume measures. *Neurology*, 72:999–1007.

Herz, R. (2007). *The Scent of Desire: Discovering our Enigmatic Sense of Smell*. New York: Harper Collins.

Herz, R.S. (1996). A comparison of olfactory, visual and tactile cues for emotional associated memories. *Chemical Senses*, 21:614–615.

Herz, R.S. (1998). Are odors the best cues to memory? A cross-modal comparison of associative memory stimuli. *Annals of the New York Academy of Sciences*, 855:670–674.

Herz, R.S. (2004). A naturalistic analysis of autobiographical memories triggered by olfactory visual and auditory stimuli. *Chemical Senses*, 29: 217–224.

Herz, R.S. and Cupchik, G.C. (1992). An experimental characterization of odor-evoked memories in humans. *Chemical Senses*, 17:519–528.

Herz, R.S., Eliassen, J., Beland, S., and Souza, T. (2004). Neuroimaging evidence for the emotional potency of odor-evoked memory. *Neuropsychologia*, 42:371–378.

Herz, R.S. and Schooler, J.W. (2002). A naturalistic study of autobiographical memories evoked by olfactory and visual cues: testing the Proustian hypothesis. *American Journal of Psychology*, 115:21–32.

Hofer, J. (1934). Medical dissertation on nostalgia. *Bulletin of the History of Medicine*, 2:376–391. (Originally published in 1688.)

Holbrook, M.B., and Schindler, R.M. (1989). Some exploratory findings on the development of musical tastes. *Journal of Consumer Research*, 16:119–124.

Hulsegge, J. and Verheul, A. (1987). *Snoezelen. Another world.* Chesterfield, UK: Rompa.

Iacoboni, M. (2008). *Mirroring People: The Science of Empathy and How We Connect with Others.* New York: Farrar Straus Giroux.

Jabbi, M., Swart M., and Keysers, C. (2007). Empathy for positive and negative emotions in the gustatory cortex. *Neuroimage*, 34:1744–1753.

Jaensch, E.R. (1930). *Eidetic Imagery*. Londen: Routledge Kegan Paul.

Kenyon, G., Bohlmeijer, E., and Randall, W.L. (eds.) (2010). *Storying Later Life. Issues, Investigations, and Interventions in Narrative Gerontology*. Oxford: Oxford University Press.

Keysers, C. (2011). *The Empathic Brain*. Groningen: Social Brain Press.

Lambrechts, D. (2003). *De smaak van heimwee: Herinneringen aan de Europese volkskeuken* [The Taste of Nostalgia: Memories of the European Folk Kitchen]. Amsterdam: Lubberhuizen.

Laird, D.A. (1935). What can you do with your nose? *Scientific Monthly*, 45:126–130.

LeDoux, J.E. (1996). *The Emotional Brain*. New York: Simon and Schuster.

LeDoux, J.E. (2007). Emotional memory. *Scholarpedia*, 2(7):1806. Available online at: <http://www.scholarpedia.org/article/Emotional_memory>.

LeDoux, J.E. (2008). Amygdala. *Scholarpedia*, 3(4):2698. Available online at: <http://www.scholarpedia.org/article/Amygdala>.

Lehrer, J. (2007). *Proust was a Neuroscientist*. Boston, MA: Houghton Mifflin Harcourt.

Lev, E. (2006). When Mozart stunned Rome; God at the pub': Wolfgang's memory caught a pope's attention'. Available online at: <http://www.zenit.org>.

Lipps, T. (1907). *Psychologie und Aesthetik* [Psychology and Aesthetics]. München.

Luria, A.R. (1968). *The Mind of a Mnemonist. A Little Book about a Vast Memory*. Translated from Russian by Lynn Solotaroff. New York/London: Basic Books.

Maguire, E.A., Gadian, D.G., Johnsrude, I.S., Good, C.D., Ashburner, J., Frackowiak, R.S., and Frith, C.D. (2000). Navigation-related structural change in the hippocampi of taxi drivers. *Proceedings of the National Academy of Sciences of the United States of America*, 97:4398–4403.

Marks, L.U. (2012). Thinking multisensory culture. In: Bacci, F. and Melcher, D. (eds.), *Art and the Senses*, pp. 239–250. Oxford: Oxford University Press.

Maurer, D. and Mondloch, C.J. (2005). Neonatal synesthesia: a reevaluation. In: Robertson, L.C. and Sagiv, N. (eds.), *Synesthesia: Perspectives from Cognitive Neuroscience*, pp. 193–213. Oxford: Oxford University Press.

McLuhan, M. (1964). *Understanding Media*. New York: McGraw Hill.

McLuhan, M. (2005). Inside the five sense sorium. In: Howes, D. (eds.), *The Empire of the Senses: The Sensual Culture Reader*, pp. 43–52. Oxford: Berg.

Meier, B. and Rothen, N. (2013). Synaesthesia and memory. In: Simner, J. and Hubbard, E. (eds.), *Oxford Handbook of Synaesthesia*. Oxford: Oxford University Press.

Merleau-Ponty, M. (2006). *Phenomenology of Perception*. Translated from French. Routledge. (Originally published in 1945.)

Moscovitch, M. (2012). The contribution of research on autobiographical memory to past and present theories of memory consolidation. In: Berntsen, D. and Rubin, D. (eds.), *Understanding Autobiographical Memory*, pp. 91–113. Cambridge: Cambridge University Press.

Moscoviter, H. (2006). *Open de luiken van uw geheugen* [Open the Shutters of your Memory]. Rotterdam: Humanitas.

Murakami, H. (2000). *Norwegian Wood*. Translated from the Japanese by Jay Rubin. Vintage. (Originally published in 1987.)

Nabokov, V. (1989). *Speak, Memory: An Autobiography Revisited*. New York: Vintage International. (Rev. edn of *Conclusive Evidence*. Originally published in 1951.)

Nabokov, V. (2002). De wandeling langs de kant van Swann [A walk along the side of Swann] In: *Een liefde voor Proust. Op zoek naar de verloren tijd in 22 leeservaringen* [A love for Proust: In Search of the Lost Time in 22 Readings]. Amsterdam: Bezige Bij.

O'Kearney, R., Speyer, J., and Kenardy, J. (2007). Children's narrative memory for accidents and their post-traumatic distress. *Applied Cognitive Psychology*, 21:821–838.

O'Keefe, J.M. and Nadel, L. (1978). *The Hippocampus as a Cognitive Map*. Oxford: Oxford University Press.

Pajo, D. (2006). *The Multi-colored Autograph and Three Versions of the Horn Concerto K495*. Available online at: <http://www.mozartforum.com/VB_forum/showthread.php?t=1888>.

Pastore, N. (1971). *Selective History of Theories of Visual Perception: 1650-1955*. New York: Oxford University Press.

Peesapati, S.T., Schwanda, V., Schultz, J., Lepage, M., Jeong, S., and Cosley, D. (2010). *Pensieve: Supporting Everyday Reminiscence*. Information Science, Cornell University. Available online at: <http://www.cs.cornell.edu/~danco/research/papers/peesapati-pensieve-chi2010.pdf>.

Ponds, R.W.H.M. (1998). *Forgetfulness and Cognitive Aging: Prevalence, Characteristics, and Determinants*. Maastricht: Universiteit Maastricht.

Pot, A.M., Bohlmeijer, E.T., Onrust, S., Melenhorst, A.S., Veerbeek, M., and De Vries, W. (2010). The impact of life review on depression in older adults: a randomized controlled trial. *International Psychogeriatrics*, 22:572–581.

Proust, M. (1917–1922/2006). *Remembrance of Things Past (In Search of Lost Time)*. Translated from French by C. K. Scott Moncrieff and Stephen Hudson. Available online at: <http://ebooks.adelaide.edu.au/p/proust/marcel/index.html>.

Purdy, D.M. (1936). Eidetic imagery and plasticity of perception. *Journal of General Psychology* 115:437–453.

Radvansky, G.A., Gibson, B.S., and McNerney, M.W. (2011). Synesthesia and memory: color congruency, von Restorff, and false memory effects. *Journal of Experimental Psychology: Learning, Memory, and Cognition*, 37:219–229.

Ramachandran, V.S. (2000). Mirror neurons and imitation learning as the driving force behind 'the great leap forward' in human evolution. Lecture 6.29.00. Available online at: <http://www.edge.org/3rd_culture/ramachandran/ramachandran_index.html>.

Ramachandran, V.S. and Hubbard, E.M. (2001). Synaesthesia—a window into perception, thought and language. *Journal of Consciousness Studies*, 8(12):3–34.

Rasmussen, A.S. and Berntsen, D. (2011). The unpredictable past: spontaneous autobiographical memories outnumber memories retrieved strategically. *Consciousness and Cognition*, 20:1842–1846.

Reese, E., Jack, F., and White, N. (2010). Origins of adolescents' autobiographical memories. *Cognitive Development*, 25:352–367.

Richardson, A. and Harris, L.J. (1986). Age trends in eidikers. *The Journal of Genetic Psychology*, 147:303–308.

Robertson, L.C. and Sagiv, N. (eds.) (2005). *Synesthesia: Perspectives from Cognitive Neuroscience*. Oxford: Oxford University Press.

Rosenblum, L.D. (2010). *See What I'm Saying: The Extraordinary Powers of Our Five Senses.* New York: Norton.

Rothen, N, and Meier, B. (2009). Do synesthetes have a general advantage in visual search and episodic memory? A case for group studies. *PLoS ONE* 4(4):e5037.

Rothen, N. and Meier, B. (2010). Grapheme-colour synaesthesia yields an ordinary rather than extraordinary memory advantage: evidence from a group study. *Memory*, 18:258–264.

Rouw, R. and Scholte, H.S. (2007). Increased structural connectivity in grapheme-color synesthesia. *Nature Neuroscience*, 10:792–797.

Rubin, D.C. (2012). The basic systems model of autobiographical memory. In: Berntsen, D. and Rubin, D. (eds.), *Understanding Autobiographical Memory*, pp. 11–32. Cambridge University Press.

Rubin, D.C. (ed.) (1988). *Autobiographical Memory*. Cambridge: Cambridge University Press.

Rubin, D.C., and Greenberg, D.L. (2003). The neuropsychology of autobiographical memory. *Cortex*, 39:687–728.

Rubin, D.C., Wetzler, S.E., and Nebes, R.D. (1986). Autobiographical memory across the adult lifespan. In Rubin, D.C. (ed.), *Autobiographical memory*, pp. 202–221. New York: Cambridge University Press.

Sacks, O. (2007). *Musicophilia. Tales of Music and The Brain*. New York: A. Knopf.

Sagiv, N., Ilbeigi, A., and Ben-Tal, O. (2011). Reflections on synesthesia, perception, and cognition. *Intellectica*, 55: 81–94

Schacter, D.L. (1996). *Searching for Memory: The Brain, the Mind and the Past.* New York: Basic Books.

Schlagman, S., Kvavilashvili, L., and Schulz, J. (2007). Effects of age on involuntary autobiographical memories. In Mace, J.H. (ed.), *Involuntary Memory*, pp. 87–112. Malden, MA: Blackwell.

Seaberg, M. (2011). *Tasting the Universe. People Who See Colors in Words and Rainbows in Symphonies*. Plompton Plains, NJ: Career Press.

Sharot, T. (2011). *The Optimism Bias: A Tour of the Irrationally Positive Brain.* New York: Pantheon.

Sharot, T., Riccardi, A.M., Raio, C.M., and Phelps, E.A. (2007). Neural mechanisms mediating optimism bias. *Nature*, 450:102.

Shulman, D. (2006). The scent of memory in Hindu South India. In Drobnik, J. (ed.), *The Smell Culture Reader*, pp. 411–426. London: Berg.

Sillevis, J. (2006). *Leven als een dandy* [Living as a Dandy]. Arnhem: Terra Lannoo.

Simner, J., Harrold, J., Creed, H., Monro, L., and Foulkes, L. (2009). Early detection of markers for synaesthesia in childhood populations. *Brain*, 132:57–64.

Simner, J. and Holenstein, E. (2007). Ordinal linguistic personification as a variant of synesthesia. *Journal of Cognitive Neuroscience*, 19:694–703.

Simner, J. and Hubbard, E.M. (2006). Variants of synesthesia interact in cognitive tasks: evidence for implicit associations and late connectivity in cross-talk theories. *Neuroscience* 143:805–814.

Simner, J. and Hubbard, E. (eds.) (2013). *Oxford Handbook of Synaesthesia*. Oxford: Oxford University Press.

Simner, J., Mayo, N., and Spiller, M.J. (2009). A foundation for savantism? Visuo-spatial synaesthetes present with cognitive benefits. *Cortex*, 45:1246–1260.

Simner, J., Mulvenna, C., Sagiv, N., Tsakanikos, E., Witherby, S.A., Fraser, C., Scott, K., and Ward, J. (2006). Synaesthesia: the prevalence of atypical cross-modal experiences. *Perception* 35:1024–1033.

Simner, J. and Ward, J. (2006). The taste of words on the tip of the tongue. *Nature*, 444:438.

Smilek, D., Malcolmson, K.A., Carriere, J.S.A., Eller, M., Kwan, D., and Reynolds, M. (2007). When '3' is a jerk and 'E' is a king: personifying inanimate objects in synesthesia. *Journal of Cognitive Neuroscience*, 19:981–992.

Smilek, D., Dixon, M.J., Cudahy, C., and Merikle, P.M. (2002). Synaesthetic color experiences influence memory. *Psychological Science*, 13:548–552.

Spector, F. and Maurer, D. (2009). Synesthesia: a new approach to understanding the development of perception. *Developmental Psychology*, 45:175–189.

Spence, C. (2011). The multisensory perception of touch. In: Bacci F. and Melcher, D. (eds.), *Art and the Senses*, pp. 85–106. Oxford: Oxford University Press.

Spence, C., Shankar, M.U., and Blumenthal, H. (2011). 'Sound bites': auditory contributions to the perception and consumption of food and drink. In Bacci F. and Mecher D. (eds.), *Art and the Senses*, pp. 207–238. Oxford: Oxford University Press.

Steunenberg, B. and Bohlmeijer, E. (2010). Life review using autobiographical retrieval: A protocol for training depressed residential home inhabitants in recalling specific personal memories. In: Kenyon, G., Bohlmeijer, E., and Randall, W.L. (eds.), *Storying Later Life. Issues, Investigations, and Interventions in Narrative Gerontology*, pp. 290–306. Oxford: Oxford University Press.

St. Jacques, P.L. (2012). Functional neuroimaging of autobiographical memory. In: Berntsen, D. and Rubin, D. (eds.), *Understanding Autobiographical Memory*, pp. 114–138. Cambridge: Cambridge University Press.

Stumpel, J. and Kieft, G. (1989). *De veelzijdige driehoek* [The Multi-sided Triangle]. *Kunstschrift*, 33(4):1–38.

Styron, W. (2001). *Darkness Visible: A Memoir of Madness.* London: Vintage. (Originally published in 1989.)

Süskind, P. (2010). *Perfume: The Story of a Murderer.* London: Penguin. (Originally published in 1985.)

Tammet, D. (2007). *Born on a Blue Day. Inside the Extraordinary Mind of an Autistic Savant.* London: Hodder & Stoughton.

Tammet, D. (2009). *Embracing the Wide Sky: A Tour Across the Horizons of the Mind.* New York: Simon & Schuster.

Tammet, D. (2012). *Thinking in Numbers: How Maths Illuminates Our Lives.* London: Hodder & Stoughton.

Toffolo, M.B., Smeets, M.A., andVan den Hout, M.A. (2011). Proust revisited: odours as triggers of aversive memories. *Cognition & Emotion* 3:1–10.

Van Abbema, D. L. and Bauer, P. J. (2005). Autobiographical memory in middle childhood: recollections of the recent and distant past. *Memory*, 13:829–845.

Verbeek, C. and Campen, C. van (2013). Inhaling memories: sense memories in art, science and practice. *Sense & Society*, 8:133–148.

Vingerhoets, A. (2013). *Why Only Humans Weep: Unravelling the Mysteries of Tears.* Oxford: Oxford University Press.

Vogelzang, M. (2008). *Eat love.* Amsterdam: BIS.

Wainwright, J. (2004). Telling times. re-visiting the greeting. In: Townsend, C. (ed.), *The Art of Bill Viola*, p. 121. London: Thames & Hudson.

Ward, J. (2008). *The Frog Who Croaked Blue: Synesthesia and the Mixing of the Senses.* Londen: Routledge.

Ward, J. and Simner, J. (2003). Lexical-gustatory synaesthesia: linguistic and conceptual factors. *Cognition*, 89:237–261.

Ward, J., Simner, J., and Auyeung, V. (2005). A comparison of lexical-gustatory and grapheme-colour synaesthesia. *Cognitive Neuropsychology*, 22:28–41.

Weert, J. van (2004) *Multi-Sensory Stimulation in 24-hour Dementia Care, Effects of Snoezelen on Residents and Caregivers.* Utrecht: NIVEL.

Werner, H. (1948). *Comparative Psychology of Mental Development.* New York: International Universities Press.

Westerhof, G.J., Lamers, S.M.A., and Vries, D.R.S.L. de (2010). Effects of recollecting autobiographical memories on the emotional well-being of older adults. *Tijdschrift voor Gerontologie en Geriatrie*, 41:5–12.

Williams, J.M.G. (1996). Depression and the specificity of autobiographical memory. In Rubin, D.C. (ed.), *Remembering Our Past: Studies in Autobiographical Memory*, pp. 244–267. Cambridge University Press.

Winocur, G., Moscovitch, M., and Bontempi, B. (2010). Memory formation and long-term retention in humans and animals: convergence towards a transformation account of hippocampal-neocortical interactions. *Neuropsychologia*, 48:2339–2356.

Wind, P. (2006). *Wondere wereld van een KeukenKoning* [The Wondrous World of a Kitchen King]. Baarn: Tirion.

Woods, B., Spector, A., Jones, C., Orrell, M., and Davies, S. (2005). Reminiscence therapy for dementia. *Cochrane Database of Systematic Reviews*, 18(2):CD001120.

Yaro, C. and Ward, J. (2007). Searching for Shereshevskii: what is superior about the memory of synaesthetes? *The Quarterly Journal of Experimental Psychology*, 60:681–695.

Yates, F.A. (1992). *The Art of Memory.* London: Pimlico. (Originally published in 1966.)

Index

Page numbers in *italics* refer to figures